State of the Art of Colloid and Interface Science in Asia

Guest Editors
To Ngai
Department of Chemistry
The Chinese University of Hong Kong
Hong Kong
China

Xiuying Qiao
School of Materials Science and Engineering
Shanghai Jiao Tong University
Shanghai
China

Editorial Office
MDPI AG
Grosspeteranlage 5
4052 Basel, Switzerland

This is a reprint of the Special Issue, published open access by the journal *Colloids and Interfaces* (ISSN 2504-5377), freely accessible at: https://www.mdpi.com/journal/colloids/special_issues/P14XH8J6BW.

For citation purposes, cite each article independently as indicated on the article page online and as indicated below:

Lastname, A.A.; Lastname, B.B. Article Title. *Journal Name* **Year**, *Volume Number*, Page Range.

ISBN 978-3-7258-7791-1 (Hbk)
ISBN 978-3-7258-7792-8 (PDF)
https://doi.org/10.3390/books978-3-7258-7792-8

Contents

About the Editors

To Ngai

To Ngai is Professor in the Department of Chemistry and Assistant Dean (Research) of the Faculty of Science at The Chinese University of Hong Kong (CUHK). He is also a Fellow of the Royal Society of Chemistry (FRSC). He received his B.Sc. in Chemistry from CUHK in 1999 and completed his Ph.D. at the same institution in 2003, focusing on light scattering and polymer interactions in solution. Following his doctorate, he joined BASF in Ludwigshafen, Germany, as a postdoctoral fellow (2003–2005), working on colloids and surface chemistry. He then undertook a short postdoctoral appointment at the University of Minnesota before returning to CUHK in 2006 as a Research Assistant Professor. He was appointed Assistant Professor in 2008, promoted to Associate Professor in 2012, and became Full Professor in 2017. His current research interests span colloids, surface chemistry, polymers, and soft matter. Prof. Ngai has published over 270 papers with >12000 citations and has an h-index of 84 (Google Scholar). He received the Chemical Society of Japan (CSJ) Lectureship Award 2020 and the Paul J. Flory Research Prize 2023 at the 29th Polychar World Forum on Advanced Materials in Nice, France.

Xiuying Qiao

Xiuying Qiao is an Associate Professor in the School of Materials Science and Engineering of Shanghai Jiao Tong University. She graduated from Changchun Institute of Applied Chemistry, Chinese Academy of Sciences in July 2001 with a Ph.D. in Polymer Chemistry and Physics. Since then, she has worked at Shanghai Jiao Tong University, first as a postdoctoral fellow in the School of Chemistry and Chemical Engineering for two years, and then in the School of Materials Science and Engineering as a Lecturer from April 2003 and as an Associate Professor from August 2005. During this time, she also worked at Kyoto University (Japan) from December 2005 to March 2007 and at the University of Akron (USA) from May 2011 to July 2012 as a postdoctoral fellow, as well as at the University of Wollongong (Australia) from June to August of 2010 as a Visiting Senior Fellow. In 2013, she won the Humboldt Fellowship for Experienced Researchers from the Alexander von Humboldt-Foundation and worked at the Max Planck Institute of Colloids and Interfaces (Germany) for over a year. Her research experience is closely related to fundamental and applied research on rheology and molecular dynamics, the interfacial stabilization and functionalization of emulsions, and the design, processing, structure, property, and application of biodegradable, biomedical, functional, smart, and high-performance polymer and polymer composites.

colloids and interfaces

MDPI

Editorial

Editorial: State of the Art of Colloid and Interface Science in Asia

To Ngai [1,*] and Xiuying Qiao [2,*]

1 Department of Chemistry, The Chinese University of Hong Kong, Hong Kong 999077, China
2 School of Materials Science and Engineering, Shanghai Jiao Tong University, Shanghai 200240, China
* Correspondence: tongai@cuhk.edu.hk (T.N.); xyqiao@sjtu.edu.cn (X.Q.)

This Special Issue brings together a diverse collection of contributions that highlight the rapid advances and emerging directions in colloid and interface science across Asia. This Special Issue comprises one mini-review and ten original research articles, reflecting the breadth of topics and the interdisciplinary nature of the field.

Colloid and interface science continues to expand its influence across energy, materials, biology, medicine, and food systems. Yet, despite remarkable progress, several gaps remain in our understanding. For example, while novel colloidal structures and functional interfaces are being developed, challenges persist in scaling fabrication methods, ensuring long-term stability, and translating laboratory findings into industrial applications. Similarly, while nanostructured materials and advanced emulsions show promise in biomedical and agricultural contexts, questions of safety, reproducibility, and environmental impact remain critical.

In the mini-review on Janus hydrogels for battery applications, Li and Wang [1] illustrate how asymmetric architectures can overcome the inherent limitations of conventional hydrogels, enabling improved ion transport and effective dendrite suppression. This work fills a critical gap in energy storage research by reframing hydrogel design through the lens of interface engineering.

The original articles span diverse applications. In the field of **nanoparticle fabrication and stabilization**, Assan et al. [2] employed a microchip laser–based pulsed ablation technique to directly generate small gold nanoparticles (~4 nm) within sensitive biopolymer matrices such as gelatin and collagen, preserving structural integrity while mitigating aggregation. Complementarily, Lei et al. [3] advanced short-wave infrared photodetectors by applying an interface engineering strategy with 3-aminopropyltriethoxysilane (APTES), which improved quantum dot–electrode contact, enhanced responsivity, controlled dark current, and accelerated photo-response.

Surface modification and characterization emerge as central themes across four studies. Huang et al. [4] demonstrated that liquid-phase plasma electrolysis (LPE) effectively removed sizing agents from carbon fiber surfaces, with SEM, TGA, and XPS confirming near-complete removal while maintaining fiber activity. In mineral processing, Xing et al. [5] showed that a blended xanthate collector (W8) with ammonium dibutyl dithiophosphate (ADD) improved gold recovery from refractory arsenopyrite ore, supported by AFM, contact angle, and adsorption density analyses. Noskov et al. [6] revealed that mixed adsorption layers of oat globulin amyloid fibrils and sodium polystyrene sulfonate formed threadlike aggregates and multilayers at liquid–gas interfaces, explaining enhanced foam stability. Shah et al. [7] systematically investigated CTAB and AOT surfactants in ethylene glycol–water mixtures, finding reduced hydrophobic interactions, looser interfacial packing, and surfactant-specific adsorption behavior, with surface tension and adhesion measurements confirming improved wettability. Collectively, these works underscore the pivotal role of surface modification and detailed characterization in optimizing material performance, flotation efficiency, interfacial stability, and wettability.

In **sensing and detection**, Meng et al. [8] reported a scalable, eco-friendly strategy for fabricating hybrid surface-enhanced Raman scattering (SERS) substrates by integrating Au nanoparticles with WO_3 nanowires via green photoreduction. The resulting Au/WO_3 nanocomposite achieved highly sensitive detection of Rhodamine 6G down to 10^{-11} M and enabled effective detection of weakly adsorbing molecules such as DMMP. In **catalysis**, Wang et al. [9] developed atomically dispersed bimetallic Pt–Sn nanocluster catalysts supported on TiO_2 to enhance hydrogenation in the toluene–methylcyclohexane cycle, a model liquid organic hydrogen carrier system. Mechanistic studies revealed that Sn incorporation modulates Pt's electronic structure, improving H_2 activation and spillover, thereby advancing liquid organic hydrogen carriers (LOHCs)-based hydrogen storage technologies.

In **biological systems and related applications**, Qiao et al. [10] highlighted silk fibroin (SF), a natural amphiphilic protein, as a versatile stabilizer of emulsions for cosmetics, food, drug delivery, and biomedicine. Nanostructured SF forms provided superior stability under harsh conditions due to strong interfacial networks. In parallel, Asmawi et al. [11] developed an eco-friendly hexaconazole-loaded nanoemulsion (Hexa-NE) to combat fungal pathogens in palm oil crops, demonstrating excellent physicochemical stability, sustained fungicide release, and highly effective antifungal activity against Ganoderma boninense.

Together, these contributions advance fundamental understanding while offering practical pathways for applying colloid and interface science to societal challenges. They address knowledge gaps in stability, scalability, and application-specific performance, while opening new opportunities for interdisciplinary collaboration. Looking forward, future research will likely emphasize the intelligent design of multifunctional colloidal systems that integrate sustainability, biocompatibility, and scalability; translational studies bridging laboratory innovation with industrial and environmental applications; and exploration of colloid science in emerging fields such as flexible electronics, biointegrated energy systems, and precision agriculture. This Special Issue underscores the vitality of colloid and interface science in Asia and beyond, inspiring further research, fostering collaboration, and stimulating industrial translation of colloid-based technologies.

Conflicts of Interest: The authors declare no conflicts of interest.

References

1. Li, P.; Wang, Q. Novel Structural Janus Hydrogels for Battery Applications: Structure Design, Properties, and Prospects. *Colloids Interfaces* **2025**, *9*, 48. [CrossRef]
2. Assan, N.; Suezawa, T.; Uetake, Y.; Yakiyama, Y.; Matsusaki, M.; Sakurai, H. Exploring the Feasibility of a Microchip Laser Ablation Method for the Preparation of Biopolymer-Stabilized Gold Nanoparticles: Case Studies with Gelatin and Collagen. *Colloids Interfaces* **2025**, *9*, 42. [CrossRef]
3. Lei, Q.; Rao, L.; Deng, W.; Ao, X.; Fang, F.; Chen, W.; Cheng, J.; Tang, H.; Hao, J. APTES-Modified Interface Optimization in PbS Quantum Dot SWIR Photodetectors and Its Influence on Optoelectronic Properties. *Colloids Interfaces* **2025**, *9*, 49. [CrossRef]
4. Huang, C.; Zhou, Q.; Li, M.; Wei, X.; Li, D.; He, X.; Chen, W. Rapid Removal of Sizing Agent from Carbon Fiber Surface by Liquid-Phase Plasma Electrolysis. *Colloids Interfaces* **2025**, *9*, 57. [CrossRef]
5. Xing, Q.; Li, F.; Ming, P.; Wang, Z. Enhanced Recovery of an Arsenopyrite-Type Gold Ore: Flotation Surface Chemistry and Kinetics of Blended Collector W8 with ADD. *Colloids Interfaces* **2025**, *9*, 76. [CrossRef]
6. Noskov, B.; Bykov, A.; Khrebina, A.; Levchuk, E.; Loglio, G.; Miller, R.; Tsyganov, E. Influence of Sodium Polystyrene Sulfonate on Surface Properties of Dispersions of Oat Globulin Fibrils. *Colloids Interfaces* **2025**, *9*, 89. [CrossRef]
7. Shah, S.; Bhattarai, R.; Gautam, S.; Shah, P.; Bhattarai, A. Wetting Behavior of Cationic and Anionic Surfactants on Hydrophobic Surfaces: Surface Tension and Contact Angle Measurements. *Colloids Interfaces* **2026**, *10*, 8. [CrossRef]
8. Meng, S.; Deng, Q.; Zhang, L.; Feng, Y.; Fan, L.; Liu, Y.; Liu, D.; Wang, C. A SERS Substrate for Ultrafast Photosynthetic Au Nanoparticle Growth on WO_3 Nanowires. *Colloids Interfaces* **2025**, *9*, 70. [CrossRef]

https://doi.org/10.3390/colloids10030034

9. Wang, J.; Lin, H.; Chan, Q.; Zhao, Y.; He, X. Atomically Dispersed Pt–Sn Nanocluster Catalysts for Enhanced Toluene Hydrogenation in LOHC Systems. *Colloids Interfaces* **2025**, *9*, 85. [CrossRef]
10. Qiao, X.; Miller, R.; Schneck, E.; Sun, K. Biocompatible Emulsions Stabilized by Natural Silk Fibroin. *Colloids Interfaces* **2026**, *10*, 13. [CrossRef]
11. Asmawi, A.; Mohd Zainudin, N.; Mohd Azman, N.; Adam, F.; Ahmad Aljafree, N.; Ahmad, M.; Abdul Rahman, M. Enhanced Lipid-Based Nanofungicide Formulation for Effective Control of Ganoderma boninense in Oil Palm. *Colloids Interfaces* **2026**, *10*, 24. [CrossRef]

https://doi.org/10.3390/colloids10030034

colloids and interfaces

MDPI

Review

Novel Structural Janus Hydrogels for Battery Applications: Structure Design, Properties, and Prospects

Ping Li and Qiushi Wang *

School of Physics and Materials Engineering, Dalian Minzu University, No. 18 Liaohe West Road, Development Area, Dalian 116600, China; liping@dlnu.edu.cn
* Correspondence: wangqiushi@dlnu.edu.cn or wangqsh6@mail.sysu.edu.cn

Abstract

Janus hydrogels, defined by their asymmetric architectures and bifunctional interfaces, have emerged as a transformative class of solid-state electrolytes in electrochemical energy storage. By integrating spatially distinct chemomechanical and ionic functionalities within a single matrix, they overcome the intrinsic limitations of conventional isotropic hydrogels, offering enhanced interfacial stability, directional ion transport, and dendrite suppression in lithium- and zinc-based batteries. This mini-review systematically highlights recent breakthroughs in Janus hydrogel design, including interfacial polymerization and layer-by-layer assembly, which collectively enable precise modulation of crosslinking gradients and ion transport pathways. This review uniquely frames Janus hydrogels from a battery-centric and interface-engineering perspective. It elucidates key structure–function correlations, identifies current limitations in scalable fabrication and electrochemical longevity, and outlines future directions toward intelligent, multifunctional platforms for next-generation flexible and biointegrated energy systems.

Keywords: Janus hydrogels; interfacial engineering; asymmetric structures; multifunctionality; batteries

1. Introduction

The continued evolution of energy storage technologies demands materials that can offer not only high ionic conductivity and electrochemical stability but also mechanical flexibility, environmental adaptability, and precise interfacial control [1–3]. Traditional hydrogel-based electrolytes—composed of hydrated polymer networks—have attracted widespread interest due to their high water content, tunable chemistry, and mechanical softness. However, most conventional hydrogels are structurally isotropic and functionally homogeneous, limiting their capacity to fulfill the complex, spatially heterogeneous requirements of modern battery architectures, particularly those involving metal anodes, flexible substrates, or multifunctional compartments [4,5].

Janus hydrogels, named after the Roman god Janus, who possesses two faces looking in opposite directions, provide a promising solution to these limitations. These hydrogels are characterized by asymmetric structures—either in chemical composition, physical morphology, or functional behavior—across distinct domains or surfaces. Inspired by biological systems such as cellular membranes or skin tissues, Janus hydrogels offer built-in spatial selectivity. This enables, for example, one side to provide ionic conduction while the other functions as a barrier layer, or one region to promote electrode contact while the other resists gas or moisture intrusion [6,7].

In energy storage systems, the dual-functional design offers multiple critical advantages by integrating complementary properties within a single structure. First, the presence of hydrophilicity gradients or asymmetrically distributed charged sites enables directional ion transport, effectively enhancing ion mobility and selectivity [8,9]. This is further complemented by selective interfacial compatibility, which facilitates improved contact with both electrodes [10] and electrolytes [11], thereby reducing interfacial resistance and promoting stable electrochemical performance. Additionally, such systems provide mechanical decoupling, allowing one layer to dissipate stress while preserving the electrochemical integrity of the other [12,13], a feature crucial for long-term cycling stability. Finally, their inherent thermal and chemical responsiveness imparts a self-regulating capability, enabling the system to adapt or shut down under abnormal conditions, thus enhancing safety and reliability [14,15].

Recent studies have demonstrated that Janus hydrogels can play multiple roles in zinc-based batteries, including functioning as quasi-solid-state electrolytes [16], artificial solid electrolyte interphase (SEI) [17], and self-healing interfaces [18]. Unlike conventional electrolytes, the asymmetric architecture of Janus hydrogels enables the spatial separation of distinct functionalities across their two faces. In lithium–metal systems, one side can be engineered to form a stable and ion-conductive artificial SEI with lithiophilic components, while the opposite side may serve as a barrier to dendritic growth, thereby achieving simultaneous stabilization of both the electrode and electrolyte interface [19]. Similarly, in aqueous zinc systems, the hydrophilic face can enhance ionic conductivity and wetting at the zinc surface, while the other face can incorporate proton-trapping or hydrogen-suppressing groups to minimize side reactions such as hydrogen evolution [20]. This spatially directed functionality not only optimizes performance in metal batteries but also makes Janus hydrogels promising for emerging applications in wearable electronics [21] and biointegrated energy systems [22], where interface specificity and multifunctionality are crucial (Figure 1).

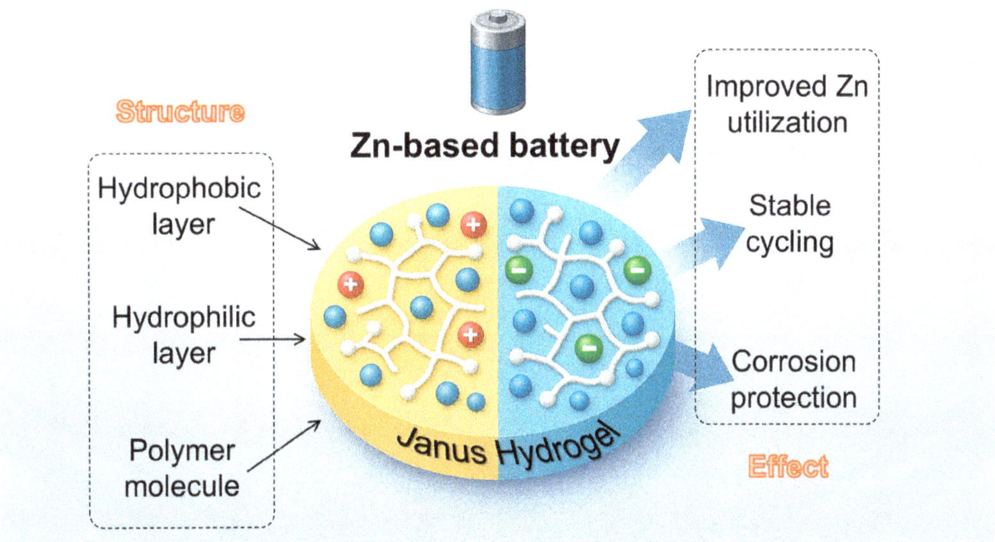

Figure 1. Schematic illustration of Janus hydrogel for Zn-based battery.

This review presents a concise yet comprehensive examination of Janus hydrogels through the lens of advanced energy storage, with an emphasis on their evolving role as intelligent, interface-engineered materials in next-generation batteries. We investigate state-of-the-art fabrication strategies, reveal the intricate structure–property–performance interplays, and assess their impact across diverse battery architectures. Special attention is given to the emergent paradigm of interfacial asymmetry as a design principle for directional ion transport, multifunctional stability, and responsive behavior. This review concludes by outlining key scientific and engineering challenges, while charting a forward-looking roadmap towards scalable manufacturing, adaptive functionality, and integration into autonomous energy systems.

2. Fabrication and Structural Features of Janus Hydrogels

The functional capabilities of Janus hydrogels stem directly from their asymmetric structure. Over the past decade, several fabrication strategies have emerged to generate spatially heterogeneous hydrogels, each offering varying degrees of morphological control, process scalability, and compatibility with electroactive materials.

2.1. Interfacial Polymerization

Interfacial polymerization is a fundamental technique in which polymer formation occurs at the interface between two immiscible or partially miscible phases. This method enables the construction of distinct layered architectures with spatially localized functionalities. For example, in a previous study, a Janus asymmetric hydrogel electrolyte (AHE) was fabricated by sequentially integrating two polymer networks with tailored roles (Figure 2a). Initially, a mixture of polyvinyl alcohol (PVA), poly(3,4-ethylenedioxythiophene)/poly(styrenesulfonate) (PEDOT:PSS), and glutaraldehyde was cast and treated with acetic acid to form a dense, conductive PVA–PEDOT hydrogel with enhanced mechanical strength on the cathode-facing side. Subsequently, a sodium alginate–carrageenan solution was layered on top and crosslinked with Zn^{2+} ions to generate a robust Carra-Zn-Alg hydrogel on the anode side. Strong hydrogen bonding at the interfacial region ensured tight adhesion and avoided delamination during swelling. This stepwise interfacial polymerization yielded a dual-layer hydrogel with asymmetric wettability and mechanical properties, significantly enhancing Zn^{2+} transport and stabilizing the Zn/electrolyte interface in Zn–I_2 batteries [23].

2.2. Gradient Curing

Gradient curing refers to a strategy that induces spatial variation in the degree of crosslinking or polymerization through differences in light exposure, initiator concentration, or component diffusion. Unlike sharply defined interfaces, this approach produces smooth, continuous gradients in regard to physical or chemical properties across the hydrogel. One notable example employed interfacial ignition driven by the density difference between a monomer precursor and a concentrated salt solution (e.g., LiCl or $ZnCl_2$). A redox-initiated polymerization front formed a barrier layer at the interface, resulting in a vertical gradient in crosslinking density—from high at the top to low at the bottom. The resulting Janus gradient hydrogel exhibited pressure-dependent conductivity, opening new possibilities for integrating sensing and energy storage functionalities [24].

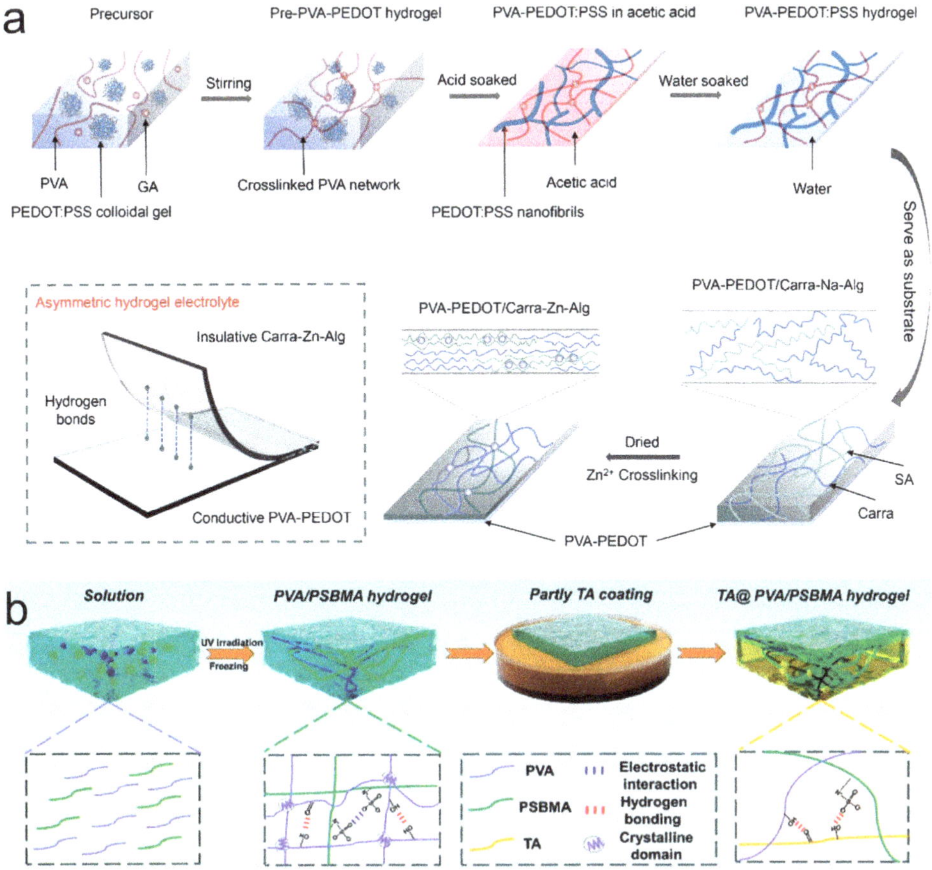

Figure 2. Strategy for fabricating a Janus adhesive hydrogel: (**a**) illustration of the preparation process for the AHE [23]; (**b**) stick–slip surface of the Janus adhesive hydrogel and its application [25].

In another case, gradient crosslinking was achieved via masked or directional ultraviolet (UV) curing, where specific regions of a Pregel matrix were selectively exposed to UV light (Figure 2b). Using this method, a Janus-structured conductive hydrogel was prepared by polymerizing sulfobetaine methacrylate (SBMA) into poly(sulfobetaine methacrylate) (PSBMA) under UV irradiation, with PVA forming a self-crosslinked backbone. Tannic acid (TA) was then introduced from one side through unidirectional dipping, forming a surface-specific modification (TA@PVA/PSBMA). The resulting hydrogel exhibited strong adhesion on the TA-modified side and excellent oil resistance on the PSBMA side, enabling antifouling and self-adhering capabilities in oil–water complex environments—ideal for wearable sensor applications [25].

2.3. Layer-by-Layer Integration

Layer-by-layer (LbL) casting is a modular technique wherein distinct hydrogel precursors are sequentially cast to create multilayered structures with spatially defined mechanical and ionic functionalities. A notable example is the development of an accordion-structured hydrogel battery (ASHB), which employs a foldable paper–gel assembly inspired by the morphology of biological organs (Figure 3a). The distinctive accordion geometry functions as an ionic isolation barrier, significantly improving the retention of ionic gradients compared to traditional laminated configurations. This structure minimizes spontaneous ion migration during storage and enables ionic gradient retention for over 30 h. The ASHB design represents a promising direction for next-generation flexible electronics by offering enhanced modularity, portability, and long-term electrochemical performance [26].

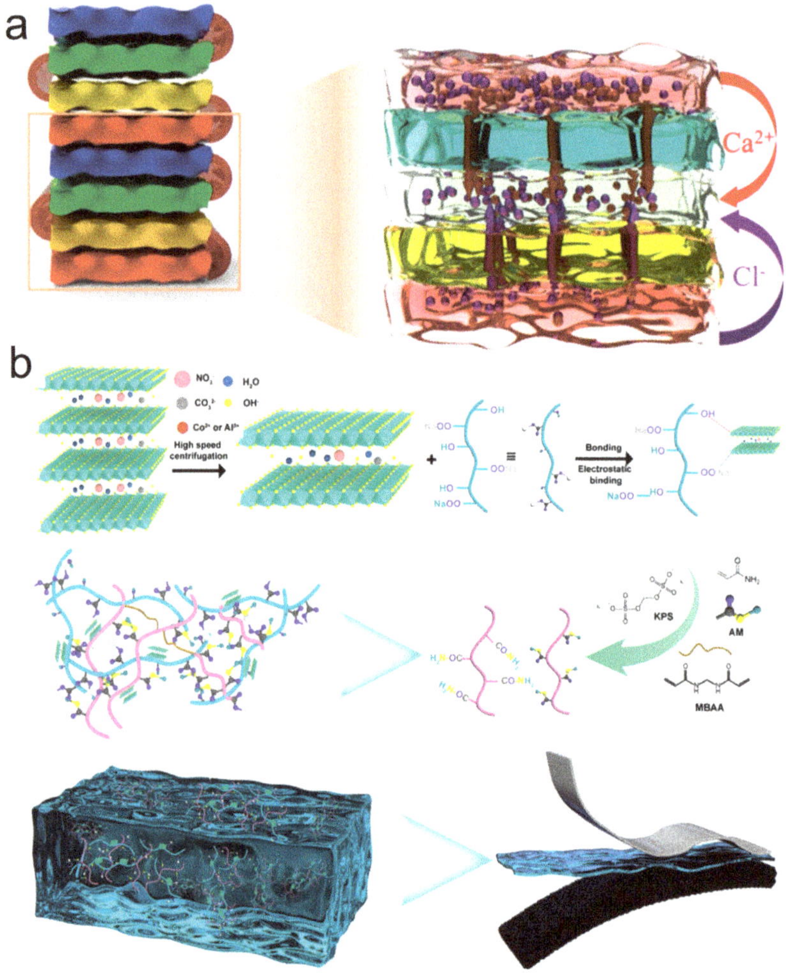

Figure 3. (**a**) Power generation principle using a paper accordion-structured hydrogel battery [26]; (**b**) CoAl LDH/CMC/PAM dual-network hydrogel preparation process diagram [27].

Further advances have leveraged sacrificial templates and anisotropic swelling agents during gelation to construct uniaxial porosity gradients. These gradients form directional ion transport channels that enhance through-thickness conductivity while minimizing lateral ionic dispersion. Although not a morphologically Janus structure in the strictest sense, such anisotropic architectures embody a core Janus design principle: directionally resolved ion transport. For instance, incorporating layered double hydroxides (LDHs) into the porous matrix yielded a dual-network hydrogel with a uniaxial pore orientation that facilitated OH^- transport through a synergistic combination of the Grotthuss and vehicle mechanisms (Figure 3b) [27]. Computational studies based on COMSOL Multiphysics simulations and activation energy analysis revealed that the Grotthuss mechanism dominated OH^- migration within the LDH-containing hydrogel, evidenced by a significantly reduced ion transport activation energy (9.14 $kJ \cdot mol^{-1}$) compared to control samples. Simulated electric field and concentration profiles further confirmed reduced internal resistance and steeper OH^- transport flux in the LDH/carboxymethyl cellulose (CMC)/polyacrylamide (PAM) system. Experimentally, this hydrogel demonstrated high ionic conductivity (145.93 $mS \cdot cm^{-1}$ at room temperature), extended battery cycling (>160 h), and excellent mechanical resilience (530% strain, 0.19 MPa stress) when used as a Zn–air battery electrolyte. These characteristics stem from anisotropic ion pathways and justify its inclusion as a Janus-inspired design.

3. Key Functional Properties of Janus Hydrogels in Battery

Janus hydrogels, by virtue of their spatially resolved architectures, exhibit a suite of physicochemical properties that directly address critical limitations in both aqueous and solid-state battery systems. Their key functionalities include directional ion transport, interfacial stabilization, mechanical adaptability, and environmental responsiveness—each arising from the rational asymmetry encoded in the hydrogel design.

3.1. Directional Ion Transport

Conventional hydrogels typically allow isotropic ion movement, which can result in uncontrolled migration, ion crossover, or uneven electrodeposition. In contrast, Janus hydrogels exploit ionic selectivity and structural anisotropy to promote unidirectional ion flux. This is especially valuable in systems prone to dendritic growth or electrolyte decomposition [28].

One common approach involves incorporating fixed charge groups—such as sulfonic acid ($-SO_3^-$) [29], carboxylate ($-COO^-$) [30], or ammonium groups—into one region of the hydrogel [31]. These fixed charges facilitate Donnan exclusion, in which counter ions (e.g., Zn^{2+}, Li^+) are preferentially absorbed and conducted while co-ions are repelled. When localized within a layered or gradient domain, this mechanism creates a chemical potential gradient that reinforces directional transport [30,32].

To overcome the limitations posed by high water reactivity in conventional aqueous and hydrogel electrolytes, particularly under elevated temperatures, a non-conventional, Janus-inspired hydrogel (HPG, a hydrogel composed of poly(ethylene glycol) methyl ether methacrylate (PMEM) and N-hydroxyethyl acrylamide (HEAA)) was rationally designed by Zhi et al. (Figure 4a) [33] This system forms a dual-functional polymer network in which the flexible ether oxygen units of PMEM coordinate Zn^{2+} transport, while the hydrophilic HEAA segments immobilize water molecules via hydrogen bonding, effectively reducing free water content and water-induced side reactions. While this hydrogel does not exhibit apparent morphological anisotropy, its molecular-level division of function between domains leads to a spatially resolved functional asymmetry. As a result, the HPG hydrogel achieves a high ionic conductivity of 3.9×10^{-3} $S \cdot cm^{-1}$ at 35 wt% water content and a significantly expanded electrochemical stability window (Figure 4b).

This asymmetrically functionalized hydrogel demonstrates remarkable temperature adaptability: symmetric Zn | | Zn and Zn | | Ti cells achieved >7500 and 5500 h of stable cycling at room temperature, and around 99% coulombic efficiency even at 90 °C, highlighting its robustness under harsh thermal conditions. In full Zn | | $Zn_{0.25}V_2O_5 \cdot nH_2O$ (ZVO) batteries, the HPG electrolyte enabled 2300 cycles with 92% capacity retention at 0.7 $A \cdot g^{-1}$, and 90% retention after 750 cycles at 90 °C with about 100% Coulombic efficiency.

These findings confirm that molecularly engineered Janus hydrogels with distinct functional domains can simultaneously ensure efficient Zn^{2+} migration and thermal resilience, offering a new strategy for high-performance, environment-adaptable aqueous zinc batteries.

3.2. Interfacial Regulation and Electrochemical Stability

The success of batteries is often governed by the quality of the electrode–electrolyte interface. In Janus hydrogels, one face can be tailored to match the chemical and mechanical characteristics of the active electrode, forming a stable, adaptive contact layer. The other face can act as a moisture barrier, electronic insulator, or an anti-gas diffusion layer [34].

In aqueous zinc–ion batteries (ZIBs), Janus hydrogels reduce hydrogen evolution reaction (HER) by spatially decoupling the Zn-plating surface from the bulk electrolyte. The hydrophobic or chemically inert outer side inhibits water penetration, while the inner layer supports ion transport and stabilizes Zn^{2+} nucleation via chelation or adsorption. Hao et al. presented a design of a Janus hydrogel electrolyte based on a combination of deep hydrophilic eutectic solvent (DES, composed of choline chloride and ethylene glycol, [ChCl]EG) and a hydrophobic ionic liquid (IL, 1-allyl-3-methylimidazolium bis(trifluoromethylsulfonyl)imide, [AMIm]TFSI), realized through a one-step UV polymerization strategy that transforms a biphasic liquid Janus system into a solid-state gel [35]. The Janus architecture comprises a hydrophilic DES phase ([ChCl]EG) and a hydrophobic IL phase ([AMIm]TFSI), connected via a miscible interfacial transition layer formed by the copolymerization of acrylic acid (AA) and trifluoroethyl methacrylate (TFEMA). This transition layer ensures structural continuity and minimal interfacial impedance between the two chemically distinct regions. In function, the DES side exhibits excellent reduction resistance due to the hydrogen bonding between choline chloride and ethylene glycol, which restricts hydrogen atom mobility. Meanwhile, the IL side offers high oxidative stability, thanks to the inherent electrochemical robustness of the $TFSI^-$ anion, providing superior compatibility with cathode materials. This dual functionality enables the Janus hydrogel to achieve a broad electrochemical stability window (ESW) of 3.6 V, along with outstanding thermal tolerance (operational from −20 to 90 °C), anti-freezing behavior (freezing point of −81.6 °C), and high solvent retention (97.47% at 25 °C over 21 days). Moreover, the hydrogel exhibits a high ionic conductivity of 3.57 $mS \cdot cm^{-1}$ and excellent long-term cycling stability, with 80.6% capacity retention and 99.9% coulombic efficiency after 10,000 charge–discharge cycles (Figure 4c,d). When used in a supercapacitor, it delivers an impressive energy density of 89.4 $Wh \cdot kg^{-1}$ at a power density of 646.2 $W \cdot kg^{-1}$.

3.3. Mechanical Modulation and Crack Suppression

Janus hydrogels frequently feature mechanical heterogeneity, where tough and stretchable components are asymmetrically distributed. This design provides built-in stress dissipation mechanisms, accommodating the swelling, shrinking, or volume changes in electrodes during cycling or thermal fluctuation.

A compelling example of utilizing structural engineering to enhance the mechanical robustness of gel-based electrolytes is demonstrated in a dual-network hydrogel system composed of poly(acrylic acid) (PAA) and PAM. Although not a classic Janus configuration in terms of compositional asymmetry, this system embodies the principle of functional compartmentalization that underpins the Janus design—achieving a balance between mechanical integrity and ionic transport through spatially differentiated structural features.

Specifically, Zhang et al. designed a PAA–PAM hydrogel that was fabricated via a stepwise crosslinking and secondary swelling strategy, forming a hierarchical pore architecture that spans from nanometer to micrometer scale [36]. X-ray tomography revealed that the micropores generated during secondary swelling act as stress-dissipating nodes, enabling the gel to withstand pressures up to 984 kPa, a record-breaking mechanical strength for hydrogel electrolytes. Simultaneously, macro-porous channels formed within the polymer networks facilitate exceptional ionic conductivity of 205 mS·cm^{-1}, which is several times higher than conventional isotropic hydrogels. Although the structure is not explicitly layered, the resulting gradient in mechanical and transport properties across the gel creates a spatially anisotropic performance profile. When applied in zinc–air battery systems—including both button cells and flexible Zn–air batteries (FZABs)—this hydrogel electrolyte supported continuous operation for over 120 h, with stable voltage output and structural integrity (Figure 4e,f).

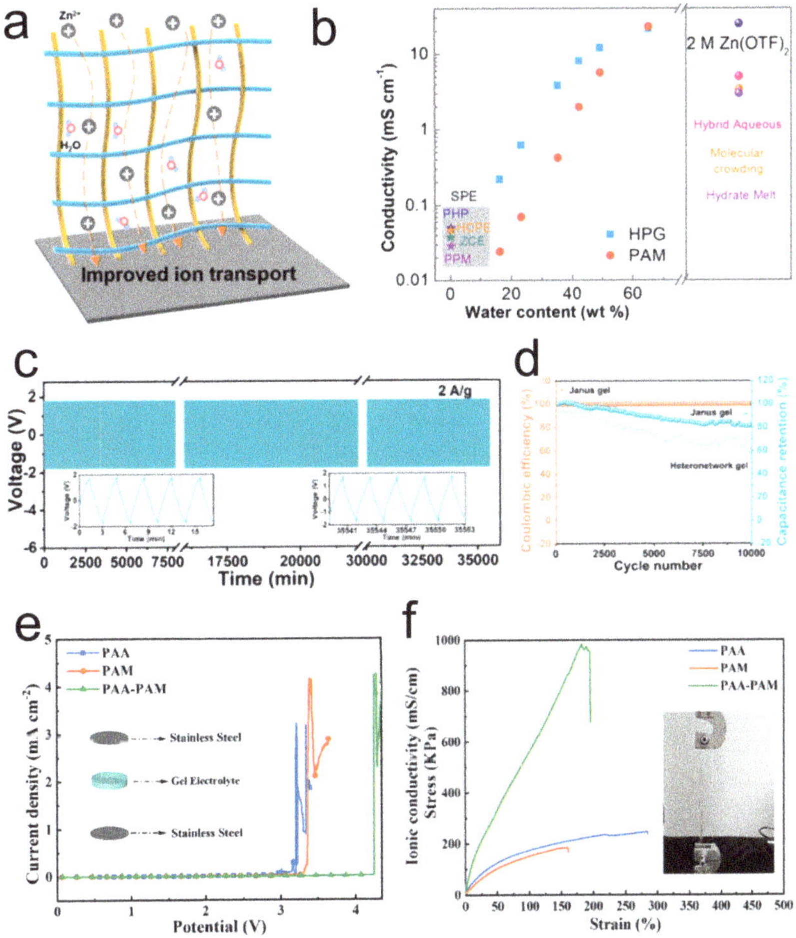

Figure 4. (**a**) Schematic of the proposed hydrogel electrolyte. (**b**) The comparison of ionic conductivity of different electrolytes: Introducing a 2,6-bis((propylimino)methyl)-4-chlorophenol (Hbimcp) ligand into the poly(propylene oxide) (PPO) polymer chain (PHP), heterolytic coordination polymer electrolytes (HCPE), Zn^{2+}-conducting solid-state electrolytes (ZCE), and precursor monomer to build in situ polymerized Zn^{2+} SPEs (PPM) [33]. (**c**) Galvanostatic charge–discharge curves of the supercapacitor at 2 A/g for 10,000 cycles. The inserts are the zoomed-in figures on the initial and end localizations of the cycles. (**d**) Cycling performance of the supercapacitors based on Janus gel and hetero-network gel as electrolytes [35]. (**e**) Electrochemical stability windows of PAA, PAM, and PAA-PAM. (**f**) Stress–strain curves of polymer gels [36].

The ability to fine-tune the pore architecture via swelling conditions offers a promising direction for the development of Janus-like hydrogels with gradient mechanical properties, where one layer may provide strain-adaptive compliance and the other reinforces load-bearing capability. This finding emphasizes the relevance of spatially organized dual-network or Janus-like hydrogels in suppressing material collapse during battery operation and provides a strategic framework for designing future electrolyte systems that simultaneously meet the demands of mechanical durability and high ionic transport.

4. Discussion

4.1. Zinc–Ion Batteries (ZIBs)

ZIBs have attracted considerable attention for grid-scale and flexible energy storage due to their intrinsic safety, low cost, and environmental friendliness. However, ZIBs are plagued by interfacial instability at the Zn anode, including uncontrolled dendrite formation, parasitic hydrogen evolution, and electrode passivation. These challenges stem from the aqueous nature of the electrolyte and the high charge density of Zn^{2+} ions. Specifically, the divalent nature and strong electrostatic interactions of Zn^{2+} lead to non-uniform ion flux and localized Zn^{2+} accumulation at the electrode interface during cycling, which promotes uneven nucleation and growth of Zn metal. This uneven deposition facilitates the formation of dendritic structures that can pierce the separator, posing serious safety and performance risks.

To address these persistent limitations, Janus hydrogel electrolytes offer a unique approach by enabling spatially resolved ion transport and interface protection within a single gel matrix. A compelling example is demonstrated in the development of rechargeable electrochromic Zn-ion batteries (RZEBs), where energy storage is coupled with dynamic color-switching functionality. Yan et al. designed a Janus gel electrolyte that was constructed by integrating a hydrophobic organogel layer composed of propylene carbonate-based poly(N,N-dimethylacrylamide) (PDMAA) and a hydrophilic hydrogel layer of PAM [37]. This asymmetric structure effectively addresses the inherent limitations of aqueous Zn systems, such as dendrite formation, HER, and Zn corrosion.

The hydrophobic PDMAA organogel, positioned adjacent to the Zn anode, acts as a physical and chemical barrier that isolates water molecules, thereby mitigating parasitic reactions and suppressing dendrite growth. Moreover, the organogel contains zincophilic sites that enable uniform Zn deposition and contribute to a stabilized Zn interface. On the other side, the PAM hydrogel provides a high-ionic-conductivity environment for the rapid Zn^{2+} insertion/extraction at the electrochromic cathode, ensuring efficient energy storage and color switching (Figure 5a). This Janus configuration results in remarkable electrochemical performance, with Zn | | Cu half-cells achieving a Coulombic efficiency of 97.91%, and Zn | | WO_3 full batteries delivering a specific capacity of 51.2 mAh·g^{-1} and maintaining 91.7% capacity retention after 100 cycles (Figure 5b). Additionally, the battery demonstrated stable and reversible optical transitions between sky-blue and black states, confirming its applicability in transparent or smart display energy systems (Figure 5c,d). This work reveals the multifunctional advantages of Janus hydrogels in aqueous Zn-ion batteries, namely simultaneously enhancing cycling stability, suppressing water-related degradation, and supporting fast ion transport, making them a powerful electrolyte design strategy for advanced multifunctional battery systems.

This example underscores how Janus hydrogels can effectively mitigate Zn anode degradation while enabling multifunctional energy storage, thus offering a strategic electrolyte design for next-generation aqueous ZIBs.

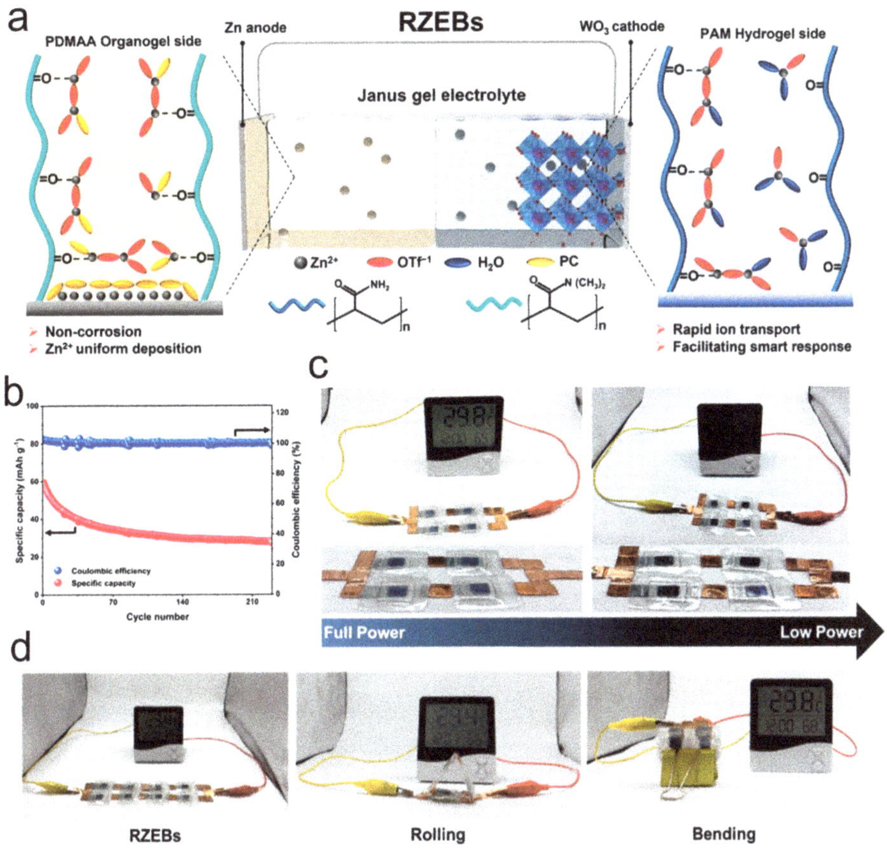

Figure 5. (**a**) Description of the preparation of Janus gel electrolyte for RZEBs; (**b**) cycling performance of RZEBs with Janus gel electrolyte at a current density of 200 mA g^{-1}; (**c**) photographs of RZEBs powering an electronic thermo-hygrometer with and without power; (**d**) images of flexible RZEBs undergoing various mechanical deformations [37].

4.2. Zinc–Air Batteries (ZABs)

ZABs are promising for high-energy-density applications due to their lightweight design and abundant raw materials. However, traditional alkaline electrolytes suffer from several drawbacks, such as rapid water evaporation, dendrite growth, Zn anode passivation, and CO_2 carbonation at the air cathode. These limitations become more severe under ambient air exposure and mechanical deformation, which are common in wearable applications.

To overcome these limitations, Fan et al. developed a Janus-type, double-layer kosmotropic gel (DLKgel), leveraging the differing kosmotropic behaviors of $ZnCl_2$ and $ZnSO_4$ salts [38]. The DLKgel comprises two functionally distinct layers within a single gel matrix: a $ZnCl_2/CaCl_2$ cellulose hydrogel positioned near the air cathode, and a $ZnSO_4$–Poly(acrylamide-acrylic acid-methylacrylamide) (PAAM) hydrogel adjacent to the Zn anode. This design enables spontaneous vertical phase separation without requiring a physical separator.

The $ZnCl_2$-rich layer promotes efficient Zn^{2+} transport and mitigates the formation of irreversible by-products such as Zn hydroxysulfates. The $ZnSO_4$ layer stabilizes the Zn anode through strong sulfate coordination, suppressing passivation and corrosion. This spatially asymmetric structure decouples ion transport functions while enhancing overall electrochemical stability. Compared to conventional alkaline or neutral gel electrolytes, the DLKgel demonstrates superior cycling stability, maintaining smooth operation for over 200 h at 0.5 mA cm^{-2} and 120 h at 1.0 mA cm^{-2} (Figure 6a).

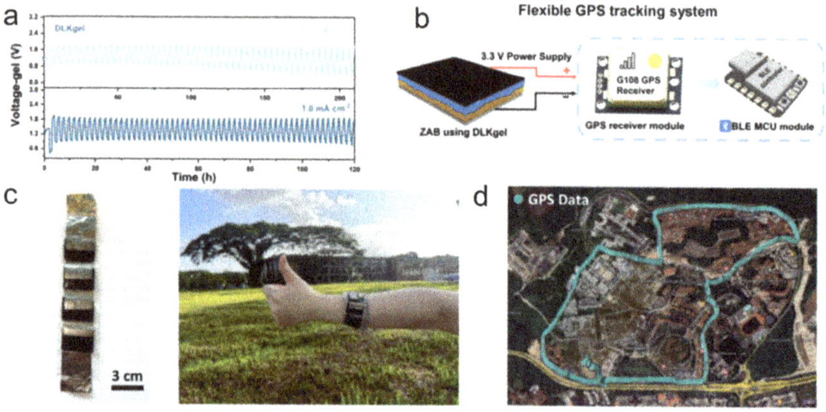

Figure 6. ZAB performance and device integration: (**a**) cycling performance of ZAB using DLKgel at two higher current densities; (**b**) schematic of the flexible GPS tracking system powered by flexible ZAB using DLKgel, including key communication protocols used to fetch data from satellites and broadcast them to mobile devices; (**c**) a wrist belt-shaped GPS recording device powered by the ZAB; (**d**) demonstration of the wearable GPS device for dynamic tracking of the trip path [38].

As a practical demonstration, the DLKgel-powered flexible ZAB was implemented in a wearable global positioning system (GPS) tracker, the battery was engineered to deliver a consistent voltage of 3.3 V, which can sustain the operation of the GPS chip (Figure 6b). These features highlight the potential of structural electrolyte engineering in extending the lifetime and practicality of ZABs for portable and wearable electronics (Figure 6c,d).

This work demonstrates how Janus hydrogel electrolytes can address the dual challenges of electrolyte degradation and mechanical adaptability, thus enabling the development of reliable, application-specific ZABs.

4.3. Flexible and Wearable Batteries

The proliferation of wearable electronics such as health monitors, smart textiles, and epidermal sensors has created an urgent need for energy storage devices that are lightweight, skin-compatible, and resilient under mechanical stress. Conventional liquid or bulk polymer electrolytes struggle to meet these criteria due to issues such as leakage, rigidity, and poor ionic interface stability.

To resolve these challenges, Yang et al. developed a Janus poly(AM-co-Zn-AC-co-NIPAM-co-MBAA) (PAZPM) hydrogel with asymmetric wettability and gradient pore structures for Zn-ion pouch cells [39]. The hydrophilic surface, with large pores, facilitates rapid proton insertion at the cathode, while the hydrophobic side, with dense pores, limits water access and stabilizes Zn deposition at the anode. This structural asymmetry balances ion accessibility with anode protection.

The Janus PAZPM hydrogel was employed in flexible Zn| |$(NH_4)_2V_{10}O_{25}{\cdot}8H_2O$ (NVO) pouch cells (Figure 7a). The resulting devices demonstrated outstanding mechanical flexibility and electrochemical performance. Pouch cells delivered a capacity of 48 mAh and retained 85% capacity after 150 cycles (Figure 7b). To highlight real-world potential, two Zn| |NVO pouch cells connected in series were shown to reliably power a 76-LED array ("DICP 504") (Figure 7c) as well as an electronic timer (Figure 7d), even under bending, cutting, or weight-loading conditions. Beyond the NVO cathode, the Janus PAZPM hydrogel also enabled stable cycling in a flexible Zn | | polyaniline (PANI) pouch cell, which delivered a high specific capacity of 154 mAh g^{-1} and 80% retention after 2800 cycles at a current density of 2 A g^{-1}, demonstrating the broad compatibility of the Janus electrolyte with different cathode chemistries.

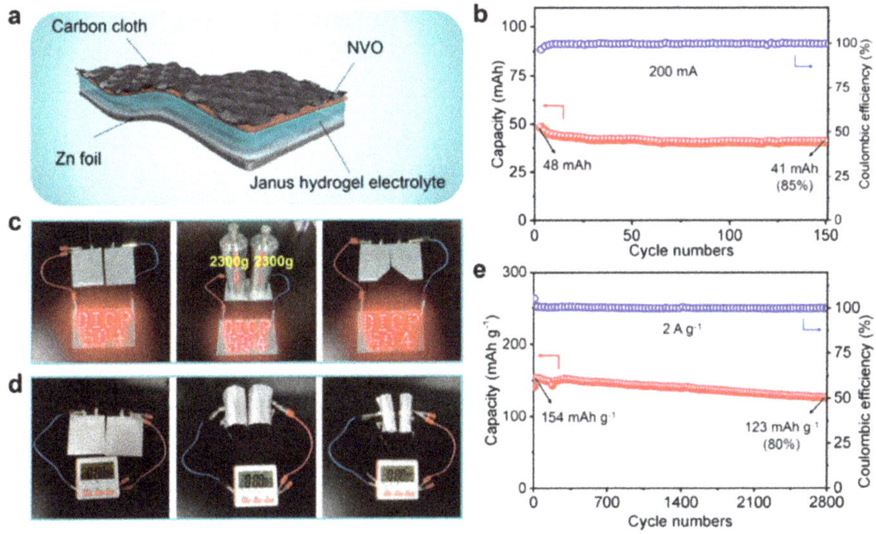

Figure 7. Demonstration of the potential of Janus PAZPM hydrogel electrolytes in flexible and wearable energy storage devices: (**a**) schematic of a flexible Zn | | NVO pouch cell employing the Janus hydrogel electrolyte; (**b**) long-term cycling stability of the Zn | | NVO pouch cell at a current of 200 mA, enabled by the Janus PAZPM hydrogel; (**c**,**d**) practical flexibility and durability of the Zn | | NVO pouch cell using the Janus hydrogel, showcased by powering LEDs and an electronic timer under bending conditions; (**e**) high-rate cycling performance of a Zn | | PANI pouch cell using the Janus PAZPM hydrogel at 2 A g^{-1}, highlighting its suitability for high-power wearable electronics [39].

This example illustrates how Janus hydrogels enable multifunctionality in wearable battery systems, offering tunable ion transport, structural robustness, and environmental adaptability. Their self-healing and stress-responsive capabilities further enhance their potential in dynamic and biocompatible energy systems.

In all these systems, Janus hydrogels play a transformative role in decoupling conflicting electrolyte functions—such as ion conductivity versus electrode protection—through spatial asymmetry. By doing so, they unlock new design pathways for high-performance, safe, and application-specific energy storage technologies. Compared to conventional liquid electrolytes, Janus hydrogels offer substantial advantages in addressing interfacial instability, dendrite formation, and electrolyte leakage. Traditional aqueous electrolytes used in ZIBs typically operate at low current densities (<1 mA·cm^{-2}) with moderate cycling lifespans (~200–500 h), whereas Janus gel-based systems demonstrate extended stability (>800 h) and high Zn utilization (>85%). In lithium–ion systems, standard carbonate-based electrolytes face challenges under mechanical strain and are prone to flammability. Solid polymer electrolytes (e.g., polyethylene oxide (PEO)-based) offer improved safety but suffer from poor room temperature conductivity (~10^{-5} S·cm^{-1}). In contrast, Janus hydrogel electrolytes can achieve ionic conductivities exceeding 10^{-3} S·cm^{-1} under ambient conditions while maintaining structural flexibility and electrochemical resilience, particularly in wearable or deformable environments. These comparisons reveal that the Janus hydrogel architecture is not merely a novel structure but addresses system-level challenges across electrolyte performance, mechanical compliance, and operational durability.

5. Challenges and Outlook

Despite the significant progress in Janus hydrogel design and the demonstrated utility of such hydrogels across various battery systems, several critical challenges must be addressed to realize their practical deployment in commercial energy storage technologies.

One major hurdle lies in the scalability and manufacturing control of current fabrication methods. The above techniques are often labor-intensive, environmentally sensitive, and unsuitable for large-area or high-throughput production. Achieving consistent batch-to-batch reproducibility—particularly in terms of interfacial alignment and crosslinking uniformity—remains difficult. To enable industrial translation, future research must focus on scalable, roll-to-roll, or printing-compatible manufacturing approaches that retain spatial precision while improving throughput and cost-efficiency.

Another pressing issue is the mechanical integrity of Janus hydrogels, particularly interfacial delamination under operational stress. Swelling mismatch, thermal gradients, or directional ion flux during cycling can weaken interfacial adhesion between asymmetrical domains, leading to performance degradation or even device failure. Enhancing interfacial robustness through strategies such as semi-interpenetrating polymer networks, dynamic covalent bonding, or modulus-gradient design is essential for ensuring mechanical durability, especially in flexible or wearable applications.

Electrochemical and environmental stability also remain limiting factors. The water-rich nature of most Janus hydrogels restricts their compatibility with high-voltage or moisture-sensitive systems such as lithium-based chemistries. Challenges such as evaporation, hydrolysis, and parasitic reactions—particularly at reactive metal interfaces—can reduce cycling life and raise safety concerns. Potential solutions include the incorporation of ionic liquids, low-volatility additives, or protective encapsulation layers to expand the electrochemical window and improve operational stability. Furthermore, the long-term biostability and ionic leakage under ambient conditions in wearable devices are still underexplored and warrant deeper investigation.

From a fundamental standpoint, the lack of comprehensive mechanistic understanding limits the rational design of Janus hydrogel systems. The impact of structural asymmetry on ion transport, charge transfer, and failure pathways remains poorly quantified. Future research should leverage in situ or operando characterization tools—such as electrochemical impedance spectroscopy, neutron scattering, and cryo-TEM—alongside multiscale computational models to elucidate structure–property–performance relationships and inform predictive design strategies.

Regarding future prospects, the next generation of Janus hydrogels is expected to incorporate adaptive, self-healing, and stimuli-responsive features. Drawing inspiration from biological systems, integrating multifunctional layers (e.g., photothermal and ion-conductive) and intelligent responsiveness (e.g., humidity-activated conductivity) will unlock broader applications beyond energy storage, including soft electronics, biosensors, and robotics. Coupling Janus hydrogels with microfluidics, soft circuitry, or AI-assisted diagnostics may ultimately enable autonomous, self-regulating electrochemical platforms for real-time monitoring and therapeutic interventions. Janus hydrogels represent more than just separators or ion conductors; they are programmable, multifunctional materials with the potential to redefine the landscape of next-generation smart electrochemical systems.

6. Conclusions

Janus hydrogels, with their asymmetric structure and dual-function interfaces, have emerged as a promising frontier in the development of next-generation solid-state electrolytes. By enabling directional ion transport, enhanced interfacial stability, and effective dendrite suppression, they provide innovative solutions to longstanding challenges in lithium-based and zinc-based batteries. This review has summarized recent progress in their structural design and fabrication, revealing how spatial control over ion dynamics and mechanical gradients can be strategically harnessed for electrochemical performance enhancement.

In terms of future prospects, Janus hydrogels hold vast potential beyond current applications. Their inherent design flexibility makes them ideal candidates for integration into a wide spectrum of emerging battery systems—including flexible and wearable electronics, implantable biomedical devices, and high-energy-density solid-state batteries. Moreover, the ability to couple stimuli-responsive, self-healing, and adaptive functionalities further positions Janus hydrogels as transformative materials in intelligent energy storage platforms. Realizing their full potential will require advances in scalable fabrication, long-term electrochemical stability, and deeper understanding of the structure–function relationship at the interface. Nonetheless, the unique convergence of multifunctionality and asymmetry in Janus hydrogels offers a powerful blueprint for engineering the next generation of smart, safe, and high-performance battery systems.

Author Contributions: Writing—original draft preparation: P.L.; writing—review and editing, funding acquisition: Q.W. All authors have read and agreed to the published version of the manuscript.

Funding: This work was supported by the National Natural Science Foundation of China (Grant Nos. 22305028, 52102211), The Science and Technique Foundation of Liaoning Province (2023JH2/101700057), and Dalian Minzu University Doctoral Program (120164, 110230).

Conflicts of Interest: There are no conflicts of interest to declare.

Abbreviations

The following abbreviations are used in this manuscript:

AA	Acrylic Acid
AHE	Asymmetric Hydrogel Electrolyte
AM	Acrylamide
[AMIm]TFSI	1-allyl-3-methylimidazolium Bis(Trifluoromethylsulfonyl)imide
ASHB	Accordion-Structured Hydrogel Battery
[ChCl]EG	Choline Chloride and Ethylene Glycol
CMC	Carboxymethyl Cellulose
DES	Deep Eutectic Solvent
DLKgel	Double-Layer Hydrogel Electrolyte
ESW	Electrochemical Stability Window
FZABs	Flexible Zn–Air Batteries
GPS	Global Positioning System
HCPE	Heterolytic Coordination Polymer Electrolytes
HEAA	N-Hydroxyethyl Acrylamide
HER	Hydrogen Evolution Reaction
HPG	Hydrogel Composed of PMEM and HEAA
IL	Ionic Liquid
LbL	Layer-by-Layer

LDH	Layered Double Hydroxides
MBAA	N,N′-methylenebisacrylamide
NIPAM	N-isopropylacrylamide
NVO	$(NH_4)_2V_{10}O_{25}\cdot 8H_2O$
PAA	Poly(Acrylic Acid)
PAAM	Poly(Acrylamide-Acrylic Acid-Methylacrylamide)
PAM	Polyacrylamide
PANI	Polyaniline
PAZPM	Poly(AM-co-Zn-AC-co-NIPAM-co-MBAA)
PDMAA	Poly(N,N-dimethylacrylamide)
PEDOT:PSS	Poly(3,4-Ethylenedioxythiophene):Poly(Styrenesulfonate)
PEO	Polyethylene Oxide
PHP	Introducing A 2,6-Bis((Propylimino)methyl)-4-Chlorophenol (Hbimcp) Ligand Into The Poly(Propylene Oxide) (PPO) Polymer Chain
PMEM	Poly(Ethylene Glycol) Methyl Ether Methacrylate
PSBMA	Poly(Sulfobetaine Methacrylate)
PVA	Polyvinyl Alcohol
RZEBs	Rechargeable Electrochromic Zn-Ion Batteries
SBMA	Sulfobetaine Methacrylate
SEI	Solid Electrolyte Interphase
semi-IPNs	Semi-Interpenetrating Polymer Networks
$-COO^-$	Carboxylate
$-SO_3^-$	Sulfonic Acid
TA	Tannic Acid
TFEMA	Trifluoroethyl Methacrylate
UV	Ultraviolet
ZABs	Zinc–Air Batteries
ZCE	Zn^{2+}-Conducting Solid-State Electrolytes
ZIBs	Zinc–Ion Batteries
Zn-AC	Zinc Acrylate
ZVO	$Zn_{0.25}V_2O_5\cdot nH_2O$

References

1. He, J.; Cao, L.; Cui, J.; Fu, G.; Jiang, R.; Xu, X.; Guan, C. Flexible energy storage devices to power the future. *Adv. Mater.* **2024**, *36*, 2306090. [CrossRef] [PubMed]
2. Dang, W.; Guo, W.; Chen, W.; Wang, J.; Zhang, Q. Extreme environment-adaptable and ultralong-life energy storage enabled by synergistic manipulation of interfacial environment and hydrogen bonding. *Energy Storage Mater.* **2025**, *74*, 103915. [CrossRef]
3. Qin, T.; Zhao, X.; Sui, Y.; Wang, D.; Chen, W.; Zhang, Y.; Luo, S.; Pan, W.; Guo, Z.; Leung, D.Y. Heterointerfaces: Unlocking superior capacity and rapid mass transfer dynamics in energy storage electrodes. *Adv. Mater.* **2024**, *36*, 2402644. [CrossRef] [PubMed]
4. Shafiei, K.; Seifi, A.; Hagh, M.T. A novel multi-objective optimization approach for resilience enhancement considering integrated energy systems with renewable energy, energy storage, energy sharing, and demand-side management. *J. Energy Storage* **2025**, *115*, 115966. [CrossRef]
5. Qorbani, M.; Chen, K.H.; Chen, L.C. Hybrid and asymmetric supercapacitors: Achieving balanced stored charge across electrode materials. *Small* **2024**, *20*, 2400558. [CrossRef] [PubMed]
6. Wu, J.; Zhang, D.; He, X.; Wang, Y.; Xiao, S.; Chen, F.; Fan, P.; Zhong, M.; Tan, J.; Yang, J. "Janus-Featured" hydrogel with antifouling and bacteria-releasing properties. *Ind. Eng. Chem. Res.* **2019**, *58*, 17792–17801. [CrossRef]
7. Luo, C.; Guo, A.; Li, J.; Tang, Z.; Luo, F. Janus hydrogel to mimic the structure and property of articular cartilage. *ACS Appl. Mater. Interf.* **2022**, *14*, 35434–35443. [CrossRef] [PubMed]
8. Guo, W.; Tian, Y.; Jiang, L. Asymmetric ion transport through ion-channel-mimetic solid-state nanopores. *Acc. Chem. Res.* **2013**, *46*, 2834–2846. [CrossRef] [PubMed]
9. Zhang, Y.; Li, F.; Kong, X.; Xue, T.; Liu, D.; Jia, P.; Wang, L.; Ding, L.; Dong, H.; Lu, D. Photoinduced directional proton transport through printed asymmetric graphene oxide superstructures: A new driving mechanism under full-area light illumination. *Adv. Funct. Mater.* **2020**, *30*, 1907549. [CrossRef]

10. Yu, X.; Wang, L.; Ma, J.; Sun, X.; Zhou, X.; Cui, G. Selectively wetted rigid–flexible coupling polymer electrolyte enabling superior stability and compatibility of high-voltage lithium metal batteries. *Adv. Energy Mater.* **2020**, *10*, 1903939. [CrossRef]
11. Zhang, K.; Yan, S.; Wu, C.; Wang, L.; Ma, C.; Ye, J.; Wu, Y. Extended battery compatibility consideration from an electrolyte perspective. *Small* **2024**, *20*, 2401857. [CrossRef] [PubMed]
12. Yuan, C.; Lu, W.; Xu, J. Electrochemical-mechanical coupling failure mechanism of composite cathode in all-solid-state batteries. *Energy Storage Mater.* **2023**, *60*, 102834. [CrossRef]
13. Song, Y.; Bhargava, B.; Stewart, D.M.; Talin, A.A.; Rubloff, G.W.; Albertus, P. Electrochemical-mechanical coupling measurements. *Joule* **2023**, *7*, 652–674. [CrossRef]
14. Zhuang, Z.; Yu, Z.; Yang, J.; Chen, L.; Xiao, T.; Fu, R.; Huang, Z.; Liu, K.; Yang, P. Thermal-gated polyanionic hydrogel films for stable and smart aqueous batteries. *Energy Storage Mater.* **2024**, *65*, 103136. [CrossRef]
15. Shokrieh, A.; Mirzaei, A.H.; Mao, L.; Shokrieh, M.M.; Wei, Z. A review of the mechanical integrity and electrochemical performance of flexible lithium-ion batteries. *Nano Res.* **2023**, *16*, 12962–12982. [CrossRef]
16. Ge, H.; Xie, X.; Xie, X.; Zhang, B.; Li, S.; Liang, S.; Lu, B.; Zhou, J. Critical challenges and solutions: Quasi-solid-state electrolytes for zinc-based batteries. *Energy Environ. Sci.* **2024**, *17*, 3270–3306. [CrossRef]
17. Han, M.; Chen, D.; Lu, Q.; Fang, G. Aqueous rechargeable Zn–iodine batteries: Issues, strategies and perspectives. *Small* **2024**, *20*, 2310293. [CrossRef] [PubMed]
18. He, X.; Zhao, L.; Zhang, Y.; Zhang, X.; Yi, J.; Xu, Q. Cathode-Hydrogel Electrolyte Interaction in Aqueous Zinc-Ion Batteries: Degradation Mechanism and Optimization Strategies. *Adv. Funct. Mater.* **2025**, e10796. [CrossRef]
19. Liu, Q.; Zhou, D.; Shanmukaraj, D.; Li, P.; Kang, F.; Li, B.; Armand, M.; Wang, G. Self-healing Janus interfaces for high-performance LAGP-based lithium metal batteries. *ACS Energy Lett.* **2020**, *5*, 1456–1464. [CrossRef]
20. Ma, H.; Chen, H.; Chen, M.; Li, A.; Han, X.; Ma, D.; Zhang, P.; Chen, J. Biomimetic and biodegradable separator with high modulus and large ionic conductivity enables dendrite-free zinc-ion batteries. *Nat. Commun.* **2025**, *16*, 1014. [CrossRef] [PubMed]
21. Zou, X.; Wang, X.; Xie, L.; Yang, Y.; Wei, C.; Zhou, Y.; Huo, P.; Yue, O.; Liu, X. Scaleable asymmetric "Janus" patch with seamless elastomer-hydrogel integration for wet on-skin electronic interfaces. *Chem. Eng. J.* **2025**, *511*, 162244. [CrossRef]
22. Zhu, T.; Wan, L.; Li, R.; Zhang, M.; Li, X.; Liu, Y.; Cai, D.; Lu, H. Janus structure hydrogels: Recent advances in synthetic strategies, biomedical microstructure and (bio) applications. *Biomater. Sci.* **2024**, *12*, 3003–3026. [CrossRef] [PubMed]
23. Liu, Q.; Yu, Z.; Fan, K.; Huang, H.; Zhang, B. Asymmetric Hydrogel Electrolyte Featuring a Customized Anode and Cathode Interfacial Chemistry for Advanced Zn–I2 Batteries. *ACS Nano* **2024**, *18*, 22484–22494. [CrossRef] [PubMed]
24. Yang, Y.; Jiang, W.; Chen, H.; Kong, F.; Wu, S.; Ni, Y.; Liu, Y. A novel strategy for preparing gradient hydrogels based on density difference-driven bidirectional self-growth. *Chem. Eng. J.* **2025**, *511*, 161936. [CrossRef]
25. Xu, L.; Wang, Y.; Li, H.; Hou, Z.; Miao, X.; Miao, G.; Li, F.; Lu, J.; Ren, G.; Zhu, X. Janus Hydrogel with Both Sticky Adhesion and Slippery Antifouling Properties for Strain Sensing. *ACS Appl. Polym. Mater.* **2024**, *6*, 2339–2348. [CrossRef]
26. Wang, J.; Huang, Y.; Gao, G.; Liu, H.; Huang, Y.; Wang, T.; Li, Z.; Shu, J.; Zhang, T. Accordion-Structured Hydrogel Battery Capable of Maintaining Ion Gradients for Extended Periods. *ACS Appl. Mater. Interf.* **2024**, *16*, 58617–58627. [CrossRef] [PubMed]
27. Yang, H.; Sun, X.; Li, X.; Liu, Q.; Hu, W.; Zhang, L. Janus Grotthuss-Vehicle Mechanism Enhances Fast OH− Transport for Ultralong Lifetime Flexible Zinc–Air Battery. *Adv. Funct. Mater.* **2024**, *34*, 2409695. [CrossRef]
28. Chen, Z.; Zhao, Y.; Cui, P.; Zhu, J.; Gao, X.; He, G.; Yi, X. Unidirectional Ion Sieve Enabling High-Flux and Reversible Zinc Anodes. *ACS Nano* **2025**, *19*, 14987–15001. [CrossRef] [PubMed]
29. Mazzaferro, L.; Grasseschi, T.M.; Like, B.D.; Panzer, M.J.; Asatekin, A. Amphiphilic Polyelectrolyte Complexes for Fouling-Resistant and Easily Tunable Membranes. *ACS Appl. Mater. Interf.* **2024**, *16*, 37952–37962. [CrossRef] [PubMed]
30. Hou, X.; Chen, X.; Liu, X.; Lu, Y.; Zou, J.; Ding, J.; Huang, K.; Xing, W.; Xu, Z. A zincophobic interface engineering achieving crystal-facet manipulation for ultra-long-life zinc-based flow batteries. *J. Membr. Sci.* **2024**, *701*, 122730. [CrossRef]
31. Jin, J.-M.; Ye, H.; Li, T.-H.; Wu, M.-B.; Liu, L.; Yao, J. Antibacterial polyamide nanofiltration membranes with intermediate layers of quaternized cellulose nanocrystal for Mg^{2+}/Li^{+} separation. *Desalination* **2024**, *582*, 117605. [CrossRef]
32. Ye, K.; Du, Y.a.; Yang, Y.; Chen, R.; Deng, C.; Weng, G.-M. An air/metal hydride battery for simultaneous neutralization treatment of acid–base wastewater and power generation. *Green Chem.* **2024**, *26*, 7891–7901. [CrossRef]
33. Wang, Y.; Liang, B.; Li, D.; Wang, Y.; Li, C.; Cui, H.; Zhang, R.; Yang, S.; Chen, Z.; Li, Q.; et al. Hydrogel electrolyte design for long-lifespan aqueous zinc batteries to realize a 99% Coulombic efficiency at 90°C. *Joule* **2025**, *9*, 101944. [CrossRef]
34. Tan, Y.; Liao, R.; Mu, Y.; Dong, L.; Chen, X.; Xue, Y.; Zheng, Z.; Wang, F.; Ni, Z.; Guo, J. Hierarchically-Structured and Mechanically-Robust Hydrogel Electrolytes for Flexible Zinc-Iodine Batteries. *Adv. Funct. Mater.* **2024**, *34*, 2407050. [CrossRef]
35. Qian, Y.; Wang, Z.; Wang, L.; Sun, X.; Qiu, X.; Hao, J. Integrated Janus gel with bilayered heterostructure for high-performance supercapacitors. *J. Colloid Interface Sci.* **2025**, *686*, 487–497. [CrossRef] [PubMed]
36. Sun, X.; Zhang, J.; Li, X.; Peng, Q.; Yang, X.; Liu, Q.; Zhang, L. Breaking the trade-off: Janus gel empowers flexible Zn-air batteries with simultaneous ultrahigh strength and ionic flux. *Chem. Eng. J.* **2025**, *514*, 163395. [CrossRef]

37. Chen, H.; Fang, P.; Yang, M.; Yu, J.; Ma, X.; Hu, Y.; Yan, F. Janus Gel Electrolyte Enabled High-Performance Quasi-Solid-State Electrochromic Zn-Ion Batteries. *ACS Appl. Polym. Mater.* **2025**, *7*, 3718–3727. [CrossRef]
38. Wu, J.; Zhang, B.; Fan, H.J. Asymmetric Kosmotropism-Stabilized Double-Layer Hydrogel for Low-Cost Neutral Zinc-Air Battery. *Small* **2024**, *20*, 2406484. [CrossRef] [PubMed]
39. Zhu, K.; Niu, X.; Xie, W.; Yang, H.; Jiang, W.; Ma, M.; Yang, W. An integrated Janus hydrogel with different hydrophilicities and gradient pore structures for high-performance zinc-ion batteries. *Energy Environ. Sci.* **2024**, *17*, 4126–4136. [CrossRef]

colloids and interfaces

MDPI

Article

Exploring the Feasibility of a Microchip Laser Ablation Method for the Preparation of Biopolymer-Stabilized Gold Nanoparticles: Case Studies with Gelatin and Collagen

Nazgul Assan [1], Tomoyuki Suezawa [1], Yuta Uetake [1,2], Yumi Yakiyama [1,2], Michiya Matsusaki [1] and Hidehiro Sakurai [1,2,*]

1 Division of Applied Chemistry, Graduate School of Engineering, The University of Osaka, 2-1 Yamadaoka, Suita 565-0871, Osaka, Japan

2 Innovative Catalysis Science Division, Institute for Open and Transdisciplinary Research Initiatives (ICS-OTRI), The University of Osaka, 2-1 Yamadaoka, Suita 565-0871, Osaka, Japan

* Correspondence: hsakurai@chem.eng.osaka-u.ac.jp; Tel.: +81-6-6879-4591

Abstract

Introducing small-sized metal nanoparticles directly into biopolymers susceptible to thermal and chemical stimulations remains a significant challenge. Recently, we showed a novel approach to fabricating gold nanoparticles through pulsed laser ablation in liquid (PLAL) using a microchip laser (MCL). Despite its lower pulse energy compared to conventional lasers, this technique demonstrates high ablation efficiency, offering the potential to produce composites without compromising the distinctive structure of biopolymers. As a proof of concept, we successfully generated gelatin-stabilized gold nanoparticles with a smaller size (average diameter of approximately 4 nm), while preserving the unchanged circular dichroism (CD) spectra, indicating the retention of gelatin's unique structure. Extending this technique to the preparation of type I collagen-stabilized gold nanoparticles yielded non-aggregated nanoparticles, although challenges in yield still persist. These results highlight the potential of the microchip laser ablation technique for producing metal nanoparticles within a vulnerable matrix.

Keywords: pulsed laser ablation in liquids (PLAL); microchip laser (MCL); gold nanoparticles; gelatin; type I collagen

1. Introduction

Incorporating small-sized metal nanoparticles (NPs) into biopolymer matrices that are susceptible to chemical and thermal stimuli remains a significant obstacle in the fields of photothermal therapy, drug delivery systems (DDS), bioimaging, and diagnosis [1–3]. Traditionally, such NPs are deposited onto biomatrices via in situ chemical methods, which often involve the use of reducing agents, pH control, temperature regulation, and other parameters [4–6]. However, these conventional chemical approaches face notable limitations, including NP agglomeration and poor dispersion, which can adversely affect the functionality of the extracellular matrix and potentially lead to its degradation.

To address these issues, a non-chemical strategy—pulsed laser ablation in liquid (PLAL)—offers several advantages for the synthesis of small-sized metal NPs directly on biopolymer matrices. Chief among these is the high purity of the resulting nanoparticles, achieved without the use of chemical additives or surfactants. Additionally, PLAL allows precise control over particle size and shape, resulting in uniform NP distribution [7,8].

Other benefits include the ease of nanoparticle functionalization and the eco-friendly nature of the process [9,10]. Notably, PLAL has been successfully applied to the synthesis of gelatin-stabilized gold nanoparticles, demonstrating its utility for generating biocompatible colloidal systems without chemical reducing agents [11]. Despite these significant advantages, there are few severe limitations, such as the stability of the biomatrix under localized heating and the potential for unwanted chemical changes due to laser irradiation. Moreover, there is the high initial cost for the laser and its setup, as well as ongoing maintenance expenses. Additionally, the yield of produced nanoparticles is lower compared to conventional chemical methods [12].

Based on this background, the PLAL method using a microchip laser (MCL) has attracted attention as a versatile approach to overcoming the limitations of conventional PLAL techniques for metal NPs synthesis. The MCL is a compact and portable device with a cavity length of only 10 mm, which can be easily implemented in a standard synthetic laboratory without the need for vibration isolation tables or complex optical systems—offering significant practical advantages [13]. Moreover, PLAL using MCL exhibits three key features that distinguish it from conventional PLAL setups. First, the system operates at a low pulse energy (1.8 mJ/pulse), minimizing the decomposition of coexisting unstable compounds and enabling the use of highly volatile solvents and additives. Second, it employs a low repetition rate (100 Hz), which, despite the low pulse power, effectively reduces the shielding effect. Third, the short pulse duration (0.9 ns) minimizes energy loss during ablation [13]. These characteristics collectively result in unique NP formation behavior.

Indeed, we previously reported that MCL-PLAL in an aqueous poly(*N*-vinyl-2-pyrrolidone) (PVP) solution consistently yields AuNPs with an average diameter of approximately 4 nm, regardless of the PVP concentration. This size uniformity is attributed to the generation of small, short-lived cavitation bubbles that govern particle formation independently of the matrix concentration [14]. Additionally, we have demonstrated the applicability of MCL-PLAL in organic solvents [15], and more recently, have shown that using aromatic solvents enables precise control over the core–shell structure of carbon-coated AuNPs [16]. These findings highlight the unique capabilities of MCL-PLAL as a platform for NP synthesis. In particular, its ability to produce uniform particles irrespective of environmental conditions and to minimize matrix damage due to low pulse energy suggests great potential for the direct introduction of AuNPs into biomatrices.

In light of this rationale, we first examined the application of MCL-PLAL to gelatin (Gel), a biopolymer previously used in PLAL-based gold nanoparticle synthesis, which typically yields particles in the 10–15 nm range, to assess the specific impact of MCL parameters. In addition, we attempted the direct synthesis of AuNPs within type I collagen (Col^{I}), a biopolymer matrix considered even more sensitive to external stimuli than gelatin.

2. Materials and Methods

Preparation of different concentrations by wt% Gel solution

Gelatin powder (077-03155, FUJIFILM Wako Pure Chemical Corp., Osaka, Japan) was dissolved in pure water at concentrations of 0.02 wt%, 0.05 wt%, 0.1 wt%, 0.2 wt%, and 0.5 wt%.

Extraction of Col^{I} from the mixture of Col^{I} and Col^{III}

A total of 1 g of collagen particles (a mixture of Col^{I} and Col^{III}, Nippon Ham, Osaka, Japan) was placed into a 2 L beaker. Then, 500 mL of water was added and stirred at 500–600 rpm in the ice bath. This mixture of collagens was stirred overnight until all the particles were dissolved. To this were added 0.45 mol/L NaCl (13.15 g) and 5 mmol/L tris-HCl buffer solution (394 mg), and the mixture was stirred for 30 min at room temperature.

Then, it was transferred to the refrigerator and stored at 4 °C overnight. Then, 1.2 mol/L NaCl (21.92 g) and 5 mmol/L tris-HCl buffer solution (394 mg) were added, and the mixture was slowly stirred for 30 min at room temperature. Once fully dissolved, it was transferred to the refrigerator and stored at 4 °C overnight. The resulting collagen solution was dispensed into conical tubes (50 mL) and centrifuged for 15 min at 10,000 rpm to afford the Col^{III} pellet (the centrifugation process should be repeated when no pellet was formed). After that, the supernatant was collected into a 2 L beaker, and the solution was transferred into dialysis membranes (15 kDa, 15 cm in length). The Col^{I} solution in membranes was immersed in water in a 2 L beaker for 7 days. On the first day, the water was changed every hour. From the second day till the seventh day, the water was changed every 3 h to afford a transparent solution. The thus-dialyzed Col^{I} solution was transferred to 50 mL conical tubes, frozen in liquid nitrogen for 30 min, and freeze-dried for 3 d at the pressure under 25 Pa.

Preparation of 0.2 wt% Col^{I} solution

The thus-extracted Col^{I} (50 mg) was placed in a 50 mL conical tube. To this was added pre-cooled PBS (25 mL), and the mixture was homogenized for 2–3 min until the sponge broke into smaller pieces. It was tightly closed and stored in the refrigerator at 4–5 °C until a homogenous solution was obtained (typically 2–3 d). Then, the solution was centrifuged under 4–5 °C for 5–7 h at 4000 rpm to remove bubbles, affording a 0.2 wt% Col^{I} solution. This solution was stored at 4–5 °C.

Procedure for PLAL method using MCL (Figure S1) for Gel/Col^{I}

A Au rod (>99.99% purity, φ 5 mm × 15 mm) was cleaned by ultrasonication in acetone for 5 min and rinsed with deionized water before use. The Au rod was fixed by a self-made holder (PEEK) and placed in a 50 mL vial (Marueme, 30 × 80 mm). Zirconia beads were placed between the holder and the sidewall of the vial to stabilize the position of the gold target during stirring and to maximize the efficiency of laser irradiation. To this was added biopolymer (Gel/Col^{I}) amount in the corresponding solvent (water/PBS, 15 mL). The laser was irradiated to the Au rod with stirring (270 rpm) for 1 h under ambient conditions in the dark. The parameters of the MCL (1064 nm, a mono-lithic Nd–YAG/Cr^{4+}– YAG ceramics) were setup on the monitor of the power supply unit (current: 70 A, pulse repetition rate: 180 Hz, pulse width: 0.9 ns). Under this setting, the laser power of 130–140 mW was obtained, which was confirmed by Nova II OPHIR Power/Energy Meter. The laser head position was adjusted to a height of 14 cm from the surface of the Au rod.

3. Results and Discussions

The primary objective of this study was to synthesize gelatin-stabilized AuNPs using MCL-PLAL without the need for a conventional chemical reducing agent. To achieve this, a bulk gold target was placed in aqueous gelatin solutions of varying concentrations (Figure 1A), and ablation was performed using an MCL. In all concentrations tested, a uniform, red-colored solution was obtained, indicating successful nanoparticle formation (Figure 1B). Watari et al. reported that when synthesizing gelatin-stabilized AuNPs via chemical reduction using sodium ascorbate, particle size varied between 10 and 15 nm depending on gelatin concentration [17]. Darroudi et al., using a high-power laser (360 mJ/pulse, 5 ns pulse duration), observed the formation of larger nanoparticles (7–19 nm), which decreased in size with longer ablation durations [11]. In this study, gelatin concentrations ranging from 0.02 to 0.5 wt% were tested under a fixed laser irradiation time of 1 h. The results are summarized in Table 1, and particle sizes were determined via Transmission Electron Microscopy (TEM) (Figure S2). The AuNPs consistently exhibited uniform sizes ranging from 4.2 to 4.4 nm and maintained a spherical morphology. A representative TEM image for the 0.2 wt% condition is shown in Figure 1C. Additionally, particle size remained constant regardless of laser irradiation time. These findings contrast sharply with previous reports. In Watari's chemical reduction method, increasing the amount of stabilizer weakened the interaction between the positively charged amino groups of gelatins and the negatively charged AuNP surfaces, leading to smaller metal core sizes. In our case, the consistent nanoparticle size is attributed to the short pulse duration of MCL-PLAL, which produces short-lived, small cavitation bubbles, thereby minimizing the impact of stabilizer concentration on particle size. Moreover, unlike in Darroudi's report, there was no evidence of secondary ablation of initially formed large particles.

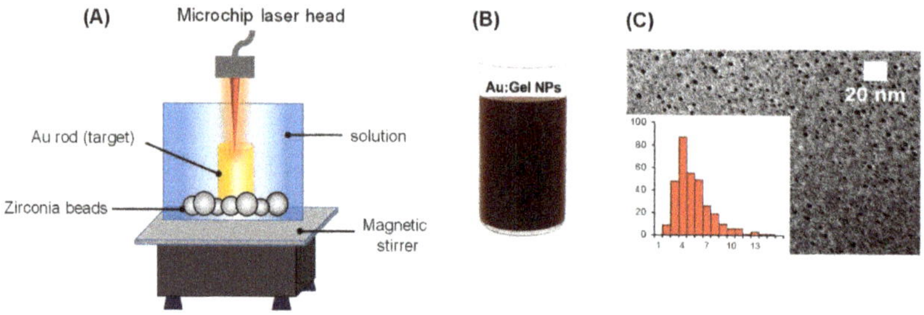

Figure 1. Setup of PLAL experiment (**A**), obtained Au NPs for Gel (**B**) and (**C**) a TEM image and the histogram in the case of entry 4, Table 1.

Table 1. Comparison of gelatin concentration, gelatin colloid size before and after formation of NPs by PLAL-MCL.

Entry	Gel Concentration (wt%)	Au–Gel NPs (nm) [a]	Yield of Au (mmol) [b]	Au–Gel Colloid Size (nm) [c]	Gel Colloid Size (nm) [c]
1	0.02	4.2 ± 2.1	4.4×10^{-5}	23 ± 4	11 ± 4
2	0.05	3.9 ± 2.0	5.7×10^{-5}	30 ± 4	7 ± 2
3	0.1	4.5 ± 2.3	2.2×10^{-5}	10 ± 4	7 ± 2
4	0.2	4.3 ± 0.9	4.4×10^{-5}	31 ± 2	7 ± 1
5	0.5	4.4 ± 2.5	7.5×10^{-5}	25 ± 3	4 ± 2

[a] measured by TEM, [b] measured by ICP-AES, [c] measured by DLS.

To investigate nanoparticle stability and potential for cellular uptake and toxicity, the colloidal size of the newly formed Au–Gel NPs was measured using dynamic light scattering (DLS). The colloidal size remained consistent across different gelatin concentrations during NP formation, except for Entry 3 (Table 1), where a significantly smaller colloidal size was consistently observed at 0.1 wt%. Pre-measurement of the gelatin colloid size prior to PLAL-MCL revealed immediate colloid formation due to gelatin's amphiphilic nature. The colloid size decreased linearly with increasing gelatin concentration. To confirm the lack of correlation between colloid size, gelatin concentration, and metal nanoparticle yield, inductively coupled plasma (ICP) measurements were conducted. The results reaffirmed that Au size, yield, and corresponding colloid sizes were independent of gelatin concentration (Table 1).

To assess structural degradation, circular dichroism (CD) spectroscopy, an important tool for obtaining information on the secondary structures of biomacromolecule [18], was performed on Au–Gel and pristine gelatin samples prepared at 0.2 wt%. Both samples showed negative peaks at 213 nm and 232 nm [19,20], characteristic of random coil structures in the polypeptide backbone. The Au–Gel sample displayed similar behavior to pristine gelatin with a negative region between 200 and 300 nm (Figure 2). The retention of negative peaks at approximately 213 nm and 232 nm in the CD spectra indicates that the random coil structure of gelatin remains largely preserved following laser ablation at 0.2 wt%, suggesting minimal disruption to its secondary structure under these conditions. CD spectra extended to 600 nm to assess chiral induction on the AuNP surface by the gelatin matrix, but no CD signal corresponding to the surface plasmon resonance (SPR) of AuNPs (500–600 nm) was observed (Figure S4). The absence of CD signals in the 500–600 nm region indicates that the surface of the AuNPs did not experience chiral induction from the surrounding matrix.

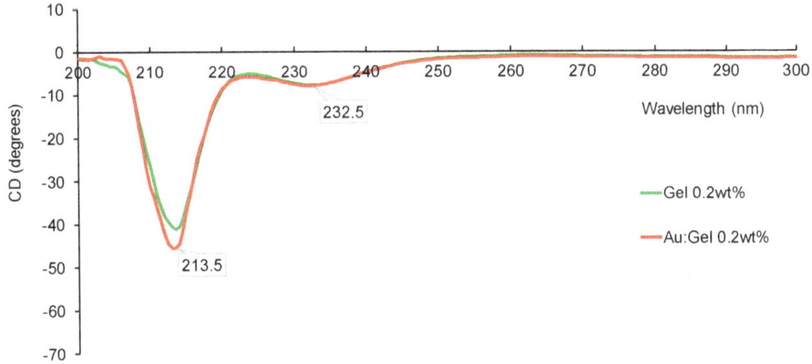

Figure 2. CD spectra of Au–Gel (red) and Gel (green) at 0.2 wt%.

In contrast, Au–Gel prepared at a lower concentration of 0.02 wt% showed different results. While gelatin exhibited two prominent negative peaks at 210 and 230 nm, more pronounced under these conditions, a new positive peak emerged around 225 nm upon laser ablation (Figure 3). This shift from the typical 232 nm signal of pristine gelatin suggests the formation of a triple helical structure, similar to partially denatured native collagen [21,22]. These findings suggest that PLAL-MCL affects the secondary structure of gelatin, promoting a more ordered conformation similar to that of collagen triple helices, which may be beneficial for materials science and biomedical applications. Collectively, these observations indicate that MCL-PLAL enables concentration-dependent fine-tuning of gelatin structure, presenting a promising strategy for designing nanocomposites with tailored bioactive conformations.

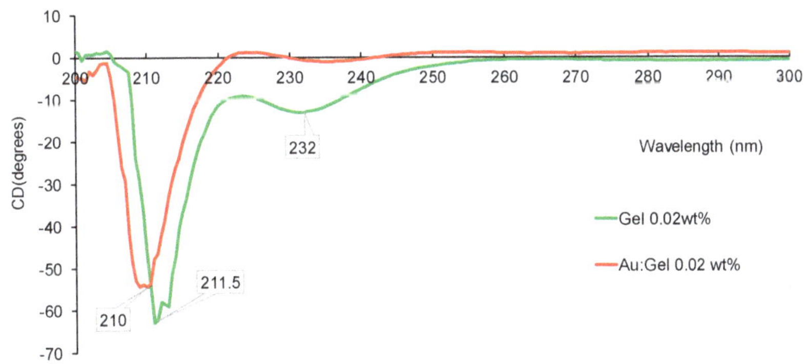

Figure 3. CD spectra of Au–Gel (red) and Gel (green) at 0.02 wt%.

Subsequent experiments investigated the influence of different media on reproducibility and Au–Gel NP synthesis. Using the optimized 0.2 wt% gelatin concentration, phosphate-buffered saline (PBS) was tested as a substitute medium to mimic cell culture conditions. The results (Table 2) showed no significant change in pH before and after ablation in pure water (6.57 → 6.81). In contrast, PBS (Entry 2) exhibited a marked increase in pH from 7.16 to 8.33 after one hour of laser irradiation. The resulting solution became visibly turbid, indicating the involvement of inorganic salts in the PBS during the ablation process (Figure S3). Laser irradiation may induce the formation of hydroxyl radicals (•OH), hydrogen atoms (•H), and dissociation of phosphate ions ($H_2PO_4^-$, HPO_4^{2-}) [23], leading to pH increase, possibly through chloride ion oxidation. Previous studies have shown that chloride ions can reduce the surface charge of AuNPs, weakening electrostatic repulsion and promoting aggregation [24,25]. Accordingly, AuNPs formed in PBS tended to aggregate over time, as stability tests revealed aggregation within one month. In contrast, those formed in water remained stable, confirming water as a more suitable medium for PLAL-based synthesis.

Table 2. Effect of different media on bio-matrix stabilized Au NPs formation by PLAL.

Entry	Matrix	Concentration (wt%)	Solvent	pH Before PLAL	pH After PLAL	Color After PLAL	ICP (mmol)
1	Gel	0.2	water	6.57	6.81	Clear red	4.4×10^{-5}
2	Gel	0.2	PBS	7.16	8.33	Cloudy red	3.8×10^{-6}
3	Col^I	0.2	PBS	6.57	6.81	Clear red	4.4×10^{-5}

Finally, we attempted direct synthesis of collagen type I (Col^I)-protected AuNPs via PLAL-MCL (Table 2, Entry 3). Unlike gelatin, Col^I cannot form stable aqueous solutions without additives, necessitating the use of PBS despite its known negative effects. After one hour of irradiation, a faint pink supernatant and red precipitate were obtained, indicating AuNP formation (Figure 4A). Centrifugation was used to separate the precipitate and supernatant for further analysis. ICP-AES revealed that 0.045 μmol of Au was ablated, with 94% present in the precipitate. Due to Col^I's high viscosity (0.082 Pa·s vs. 0.001 Pa·s for water), ablation efficiency was reduced, resulting in lower AuNP yield [26–28].

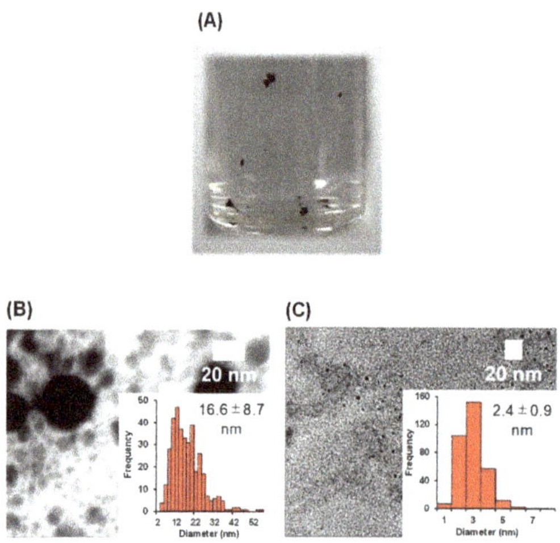

Figure 4. (**A**) Au–Col^I dispersion after PLAL showing precipitate and supernatant. (**B**) TEM and size distribution of nanoparticles from the precipitate (18.6 ± 8.7 nm). (**C**) TEM and size distribution from the supernatant (2.4 ± 0.9 nm). Scale bars: 20 nm.

TEM analysis of the red precipitate confirmed AuNP formation with a mean diameter of 16.6 ± 8.7 nm (Figure 4B). However, the precipitates could not be redispersed in any aqueous solution, indicating that significant deformation of Col^{I} might occur during the ablation process. On the other hand, the supernatant, containing 6% of the total Au, showed much smaller and more uniform AuNPs (2.4 ± 0.9 nm) (Figure 4C), the smallest known for Col^{I} matrices [28].

4. Conclusions

In summary, we successfully established a method for the direct incorporation of AuNPs into highly susceptible biopolymer matrices using pulsed laser ablation in liquid (PLAL) with a microchip laser (MCL), without relying on any chemical reduction processes. Despite the low pulse energy of the laser, the method achieved high efficiency. This approach enabled the formation of gelatin-stabilized AuNPs with an average size of approximately 4 nm, without altering the structural integrity of the gelatin—a result that is difficult to achieve with conventional techniques. Moreover, when gelatin was used at low concentrations, there was an indication that the laser process may have influenced the secondary structure of the gelatin.

On the other hand, when the method was extended to type I collagen—an even more sensitive biomatrix—a majority of the gold nanoparticles were entrapped within the aggregated matrix. Nevertheless, we successfully synthesized a small fraction of well-dispersed, non-aggregated collagen-stabilized AuNPs of small size. One possible reason for the low yield in this case may be the degradation of the PBS buffer caused by laser irradiation. If this is indeed the case, the use of PBS may need to be avoided, potentially limiting the applicability of this method to certain biomatrices. Nonetheless, the MCL-PLAL method presented in this study offers a fundamentally distinct and meaningful approach compared to conventional techniques, particularly in enabling NP synthesis under mild conditions within sensitive biological environments.

Supplementary Materials: The following supporting information can be downloaded at: https://www.mdpi.com/article/10.3390/colloids9040042/s1, Figure S1: Photographs of the experimental setup used for microchip laser ablation; Figure S2: TEM images and size distribution histograms of Au:Gel nanoparticles at varying gelatin loadings: 0.02% (A), 0.05% (B), 0.1% (C), and 0.5% (D). Scale bars: 20 nm; Figure S3: Transmission electron microscopy (TEM) images of Au:Gel nanoparticles at low (200 nm) and high (20 nm) magnification, along with corresponding particle size distribution histogram; Figure S4. Circular dichroism (CD) spectra of Au:Gel nanoparticles at 0.02 wt% (red) and 0.2 wt% (purple) gelatin concentrations. Table S1: Zeta potential measurements for Au:Gel (main experiment) and Gel (control experiment).

Author Contributions: Conceptualization, H.S.; methodology, N.A. and T.S.; validation, Y.U., Y.Y., M.M. and H.S.; formal analysis, N.A. and T.S.; investigation, N.A. and T.S.; writing—original draft preparation, N.A.; writing—review and editing, N.A., Y.U., Y.Y. and H.S.; supervision, M.M. and H.S.; project administration, H.S. All authors have read and agreed to the published version of the manuscript.

Funding: JSPS KAKENHI grant JP19K22187, JP24H00460 (H.S.); JP20K15279 (Y.U.); COI-NEXT (JPMJPF2009) from JST (M.M.); JPNP20004 from NEDO (M.M.), the Amada Foundation for the Promotion of Science & Engineering (AF2018234-C2) (Y.Y.), and Special Project by Institute for Molecular Science (IMS program 21-259) (H.S. and Y.Y.).

Data Availability Statement: Data is contained within the article or Supplementary Materials.

Acknowledgments: N.A. acknowledges MEXT for the scholarship.

Conflicts of Interest: The authors declare no conflicts of interest.

References

1. Kogan, M.J.; Olmedo, I.; Hosta, L.; Guerrero, A.R.; Cruz, L.J.; Albericio, F. Peptides and Metallic Nanoparticles for Biomedical Applications. *Nanomedicine* **2007**, *2*, 287–306. [CrossRef] [PubMed]
2. Khlebtsov, N.G.; Dykman, L.A. Optical Properties and Biomedical Applications of Plasmonic Nanoparticles. *J. Quant. Spectrosc. Radiat. Transf.* **2010**, *111*, 1–35. [CrossRef]
3. Elahi, N.; Kamali, M.; Baghersad, M.H. Recent Biomedical Applications of Gold Nanoparticles: A Review. *Talanta* **2018**, *184*, 537–556. [CrossRef] [PubMed]
4. Chatterjee, S.; Lou, X.Y.; Liang, F.; Yang, Y.W. Surface-Functionalized Gold and Silver Nanoparticles for Colorimetric and Fluorescent Sensing of Metal Ions and Biomolecules. *Coord. Chem. Rev.* **2022**, *459*, 214461. [CrossRef]
5. Fatima, Z.; Saleem, R.; Khan, R.R.M.; Liaqat, M.; Pervaiz, M.; Saeed, Z.; Muhammad, G.; Amin, M.; Rasheed, S. Green Synthesis, Properties, and Biomedical Potential of Gold Nanoparticles: A Comprehensive Review. *Biocatal. Agric. Biotechnol.* **2024**, *59*, 103271. [CrossRef]
6. Liu, X.; Luo, Y.; Zhang, Y.; Xie, Z.; Xu, C. Gold Nanoparticle-Mediated Fluorescence Resonance Energy Transfer for Analytical Applications in the Fields of Life Health and Safety. *Talanta* **2025**, *282*, 127023. [CrossRef]
7. Kim, M.; Osone, S.; Kim, T.; Higashi, H.; Seto, T. Synthesis of Nanoparticles by Laser Ablation: A Review. *KONA Powder Part. J.* **2017**, *2017*, 80–90. [CrossRef]
8. Balachandran, A.; Sreenilayam, S.P.; Madanan, K.; Thomas, S.; Brabazon, D. Nanoparticle Production via Laser Ablation Synthesis in Solution Method and Printed Electronic Application—A Brief Review. *Results Eng.* **2022**, *16*, 100646. [CrossRef]
9. He, Y.; Wang, L.; Wu, T.; Wu, Z.; Chen, Y.; Yin, K. Facile Fabrication of Hierarchical Textures for Substrate-Independent and Durable Superhydrophobic Surfaces. *Nanoscale* **2022**, *14*, 9392–9400. [CrossRef]
10. Yin, K.; Wang, L.; Deng, Q.; Huang, Q.; Jiang, J.; Li, G.; He, J. Femtosecond Laser Thermal Accumulation-Triggered Micro-/Nanostructures with Patternable and Controllable Wettability Towards Liquid Manipulating. *Nano-Micro Lett.* **2022**, *14*, 97. [CrossRef]
11. Darroudi, M.; Khorsand Zak, A.; Muhamad, M.R.; Zamiri, R. Preparation of Gelatinous Gold Nanoparticles by Pulsed Laser Ablation. *Res. Chem. Intermed.* **2015**, *41*, 4587–4594. [CrossRef]
12. Jiang, Z.; Li, L.; Huang, H.; He, W.; Ming, W. Progress in Laser Ablation and Biological Synthesis Processes: "Top-Down" and "Bottom-Up" Approaches for the Green Synthesis of Au/Ag Nanoparticles. *Int. J. Mol. Sci.* **2022**, *23*, 14658. [CrossRef] [PubMed]
13. Lim, H.H.; Taira, T. Sub-Nanosecond Laser Induced Air-Breakdown with Giant-Pulse Duration Tuned Nd:YAG Ceramic Micro-Laser by Cavity-Length Control. *Opt. Express* **2017**, *25*, 6302. [CrossRef] [PubMed]
14. Hettiarachchi, B.S.; Takaoka, Y.; Uetake, Y.; Yakiyama, Y.; Lim, H.H.; Taira, T.; Maruyama, M.; Mori, Y.; Yoshikawa, H.Y.; Sakurai, H. Uncovering gold nanoparticle synthesis using a microchip laser system through pulsed laser ablation in aqueous solution. *Ind. Chem. Mater.* **2024**, *2*, 340–347. [CrossRef]
15. Hettiarachchi, B.S.; Takaoka, Y.; Uetake, Y.; Yakiyama, Y.; Yoshikawa, H.Y.; Maruyama, M.; Sakurai, H. Mechanistic Study in Gold Nanoparticle Synthesis through Microchip Laser Ablation in Organic Solvents. *Metals* **2024**, *14*, 155. [CrossRef]
16. Hettiarachchi, B.S.; Yakiyama, Y.; Sakurai, H. Tailoring Carbon-Encapsulated Gold Nanoclusters via Microchip Laser Ablation in Polystyrene Solution: Controlling Size, Structure, and Photoluminescent Properties. *RSC Appl. Interfaces* **2025**, *2*, 772–779. [CrossRef]
17. Neupane, M.P.; Lee, S.J.; Park, I.S.; Lee, M.H.; Bae, T.S.; Kuboki, Y.; Uo, M.; Watari, F. Synthesis of Gelatin-Capped Gold Nanoparticles with Variable Gelatin Concentration. *J. Nanopart. Res.* **2011**, *13*, 491–498. [CrossRef]
18. Miles, A.J.; Janes, R.W.; Wallace, B.A. Tools and methods for circular dichroism spectroscopy of proteins: A tutorial review. *Chem. Soc. Rev.* **2021**, *50*, 8400–8413. [CrossRef]
19. Gopal, R.; Park, J.S.; Seo, C.H.; Park, Y. Applications of Circular Dichroism for Structural Analysis of Gelatin and Antimicrobial Peptides. *Int. J. Mol. Sci.* **2012**, *13*, 3229–3244. [CrossRef]
20. Rub, M.A.; Asiri, A.M.; Khan, J.M.; Khan, R.H.; Din, K.-U. Interaction of Gelatin with Promethazine Hydrochloride: Conductimetry, Tensiometry and Circular Dichroism Studies. *J. Mol. Struct.* **2013**, *1050*, 35–42. [CrossRef]
21. Asil, S.M.; Narayan, M. Molecular Interactions between Gelatin-Derived Carbon Quantum Dots and Apo-Myoglobin: Implications for Carbon Nanomaterial Frameworks. *Int. J. Biol. Macromol.* **2024**, *264*, 130416. [CrossRef] [PubMed]
22. Lee, S.; Jo, K.; Kim, S.; Woo, M.; Choi, Y.; Jung, S. Food Hydrocolloids. *Food Hydrocoll.* **2025**, *160*, 110739. [CrossRef]
23. Correard, F.; Maximova, K.; Estève, M.A.; Villard, C.; Roy, M.; Al-Kattan, A.; Sentis, M.; Gingras, M.; Kabashin, A.V.; Braguer, D. Gold Nanoparticles Prepared by Laser Ablation in Aqueous Biocompatible Solutions: Assessment of Safety and Biological Identity for Nanomedicine Applications. *Int. J. Nanomed.* **2014**, *9*, 5415–5430. [CrossRef]
24. Lévy, A.; De Anda Villa, M.; Laurens, G.; Blanchet, V.; Bozek, J.; Gaudin, J.; Lamour, E.; Macé, S.; Mignon, P.; Milosavljević, A.R.; et al. Surface Chemistry of Gold Nanoparticles Produced by Laser Ablation in Pure and Saline Water. *Langmuir* **2021**, *37*, 5783–5794. [CrossRef]

25. Kwon, H.; Kim, K.K.; Song, J.K.; Park, S.M. The Effects of Ambient Ions on the Growth of Gold Nanoparticles by Laser Ablation in Liquid. *Bull. Korean Chem. Soc.* **2014**, *35*, 865–870. [CrossRef]
26. Costela, A.; Garcia-Moreno, I.; Barroso, J.; Sastre, R. Laser Performance of Coumarin 540A Dye Molecules in Polymeric Host Media with Different Viscosities: From Liquid Solution to Solid Polymer Matrix. *J. Appl. Phys.* **1998**, *83*, 650–660. [CrossRef]
27. Takata, T.; Enoki, M.; Chivavibul, P.; Matsui, A.; Kobayashi, Y. Effect of Confinement Layer on Laser Ablation and Cavitation Bubble during Laser Shock Peening. *Mater. Trans.* **2016**, *57*, 1776–1783. [CrossRef]
28. Hupfeld, T.; Laurens, G.; Merabia, S.; Barcikowski, S.; Gökce, B.; Amans, D. Dynamics of Laser-Induced Cavitation Bubbles at a Solid-Liquid Interface in High Viscosity and High Capillary Number Regimes. *J. Appl. Phys.* **2020**, *127*, 044306. [CrossRef]

colloids and interfaces

MDPI

Article

APTES-Modified Interface Optimization in PbS Quantum Dot SWIR Photodetectors and Its Influence on Optoelectronic Properties

Qian Lei [1,2,†], Lei Rao [1,†], Wencan Deng [1], Xiuqin Ao [1,2], Fan Fang [3], Wei Chen [3], Jiaji Cheng [2], Haodong Tang [1,*] and Junjie Hao [1,*]

[1] College of Integrated Circuits and Optoelectronic Chips, Shenzhen Technology University, Shenzhen 518118, China; 202321113012284@stu.hubu.edu.cn (Q.L.); 202201102083@stumail.sztu.edu.cn (L.R.); 202100303020@stumail.sztu.edu.cn (W.D.); 202321113012491@stu.hubu.edu.cn (X.A.)

[2] School of Materials Science and Engineering, Hubei University, Wuhan 430062, China; jiajicheng@hubu.edu.cn

[3] College of Engineering Physics, and Center for Intense Laser Application Technology, Shenzhen Technology University, Shenzhen 518118, China; fangfan@sztu.edu.cn (F.F.); chenwei@sztu.edu.cn (W.C.)

* Correspondence: tanghaodong@sztu.edu.cn (H.T.); haojunjie@sztu.edu.cn (J.H.)

† These authors contributed equally to this work.

Abstract

Lead sulfide colloidal quantum dots (PbS QDs) have demonstrated great potential in short-wave infrared (SWIR) photodetectors due to their tunable bandgap, low cost, and broad spectral response. While significant progress has been made in surface ligand modification and defect state passivation, studies focusing on the interface between QDs and electrodes remain limited, which hinders further improvement in device performance. In this work, we propose an interface engineering strategy based on 3-aminopropyltriethoxysilane (APTES) to enhance the interfacial contact between PbS QD films and ITO interdigitated electrodes, thereby significantly boosting the overall performance of SWIR photodetectors. Experimental results demonstrate that the optimal 0.5 h APTES treatment duration significantly enhances responsivity by achieving balanced interface passivation and charge carrier transport. Moreover, The APTES-modified device exhibits a controllable dark current and faster photo-response under 1310 nm illumination. This interface engineering approach provides an effective pathway for the development of high-performance PbS QD-based SWIR photodetectors, with promising applications in infrared imaging, spectroscopy, and optical communication.

Keywords: photodetector; PbS; APTES

1. Introduction

Colloidal quantum dots (QDs), particularly lead sulfide (PbS) QDs [1–3], owing to their tunable bandgap, simple synthesis process, and low cost, exhibit broad spectral tunability in the short-wave infrared (SWIR) range. They have been widely applied in areas such as infrared detection [4–7], solar cells [8–10], and biomedical sensing [11–13]. Among these, PbS QDs are especially prominent in SWIR photodetectors, primarily because of their low cost and solution processability. In addition, their high dielectric constant, long carrier lifetime, and the potential for multiple-exciton generation contribute to enhanced light absorption and carrier collection efficiencies. More importantly, their process compatibility with large-area substrates and monolithic integration with CMOS circuits [14,15] make them ideal for developing high-resolution, low-crosstalk imaging arrays.

Despite these advantages, a major challenge for PbS QD-based photodetectors lies in the poor charge transport between adjacent QDs, primarily due to long insulating ligands such as oleic acid (OA) used during synthesis [16,17]. These long ligands act as tunneling barriers, limiting carrier mobility and reducing device responsivity. To address this, several strategies have been explored. One effective approach involves employing a two-step ligand-exchange method, which reduces the inter-dot spacing and increases the ligand exchange efficiency. This enhancement in charge transfer capability led to a significant 94% improvement in photodetector responsivity [18]. Additionally, optimizing QD synthesis via the perovskite conversion method (PCM) further facilitates charge carrier transport within the device, yielding additional responsivity improvements [19]. However, investigations into the interface between QDs and electrodes remain limited, despite its critical role in further enhancing device performance.

The interaction between silane coupling agents and QDs has been extensively studied [20–22]. In this work, we present a novel and effective strategy for enhancing the performance of PbS QD SWIR photodetectors via interfacial engineering using 3-aminopropyltriethoxysilane (APTES). By modifying the surface of ITO interdigitated electrodes with an APTES layer, the interfacial contact and film morphology between the QD layer and the electrode are significantly enhanced [23–27]. This modification promotes more uniform and denser QD films, facilitates charge transport, and suppresses defect-induced recombination. Systematic studies reveal that the APTES treatment duration is a critical parameter: insufficient treatment results in incomplete passivation, whereas excessive treatment introduces additional tunneling barriers and increases series resistance. An optimized treatment time of 0.5 h achieves the best balance between interface passivation and carrier extraction, resulting in markedly improved responsivity and reduced dark current. These findings not only offer a practical solution for optimizing QD-based photodetectors but also provide new insights into interface engineering for next-generation optoelectronic devices.

2. Materials and Methods

2.1. Materials

Lead (II) chloride ($PbCl_2$, ≥99.99%), sublimed sulfur (S, ≥99.5%), and tetrabutylammonium iodide (TBAI, ≥99%) were purchased from Aladdin (Shanghai, China). 3-Aminopropyltriethoxysilane (APTES, 99%), 1-octadecene (ODE, 90.0%, GC), and n-octane (98%) were obtained from Macklin. Oleic acid (OA, 99%), oleylamine (OlAm, C18 content: 80–90%), n-hexane (≥99.5%), ethanol (≥99.5%), and methanol (≥99.9%) were purchased from Energy Chemical (Anqing, China). All chemicals were used as received without further purification.

2.2. Synthesis of PbS QDs

S precursor: In a three-neck flask, 0.16 g (5 mmol) of sulfur (S) was dissolved in 15 mL of oleylamine (OlAm). The flask was subjected to three cycles of evacuation and nitrogen purging. Then, vacuum was slowly applied until no bubbles were observed. When the temperature reached 120 °C, the system was switched to nitrogen atmosphere and heated for 30 min.

Pb precursor: In another three-neck flask, 0.834 g (3 mmol) of lead (II) chloride ($PbCl_2$) and 7.5 mL of OlAm were mixed. The mixture underwent three cycles of evacuation and nitrogen purging and then was slowly evacuated until no bubbles appeared. Once the temperature reached 125 °C, the system was switched to nitrogen and heated for 30 min.

At 120 °C, 2.25 mL of the S-OlAm solution (containing 0.75 mmol of sulfur) was swiftly injected into the Pb precursor solution. The reaction was allowed to proceed for 170 s. After the reaction, 10 mL of n-hexane and 15 mL of ethanol were added to quench the reaction.

2.3. Ligand Exchange of PbS QDs

The product was collected by centrifugation at 3000 rpm for 3 min, and the supernatant was discarded. The precipitated QDs were redispersed in 10 mL of n-hexane, followed by the addition of 1.5 mL of OA. After thorough mixing, 15 mL of ethanol was added, and the mixture was centrifuged at 3000 rpm for 3 min. The precipitate was redispersed in 10 mL of n-hexane and then reprecipitated with 15 mL of ethanol and centrifuged again at 3000 rpm for 3 min. This washing process was repeated twice. Finally, the precipitate was dried under vacuum for 10 min and stored [28,29].

2.4. Preparation of APTES-Modified SWIR Photodetectors

APTES precursor: A 0.05 M APTES solution in n-hexane was prepared by mixing 217.5 μL of APTES with 49.7825 mL of n-hexane to make a total volume of 50 mL.

PbS QDs stock solution: PbS QDs were dispersed in n-octane to obtain a concentration of 50 mg/mL. The resulting solution was filtered twice through a 0.25 μm polytetrafluoroethylene (PTFE) membrane to ensure uniformity and purity.

ITO substrate cleaning: ITO substrates were cleaned using a glass cleaner diluted with water (1:4, v/v), followed by sequential ultrasonication in deionized water, isopropanol, acetone, and ethanol for 20 min each. After cleaning, the substrates were treated with ozone for 15 min to remove any residual organic contaminants.

APTES modification of electrodes: ITO interdigitated electrodes were immersed into the APTES precursor solution for a defined period and then rinsed thoroughly with n-hexane to remove unbound APTES. The immersion time was varied to control the degree of functionalization and thereby tune the interaction between the electrode surface and the PbS QDs.

QDs layer deposition: The 50 mg/mL PbS QD stock solution in n-octane was spin-coated onto the substrate at 3000 rpm for 30 s to form the QD film. A 10 mg/mL solution of tetrabutylammonium iodide (TBAI) in methanol was then drop-cast onto the film and allowed to sit for 30 s. This was followed by spin-coating at 4000 rpm for 10 s. Methanol was applied with a pipette to fully cover the surface, followed by spin-coating at 4000 rpm for another 10 s. The methanol washing step was repeated three times. Steps involving QD deposition, TBAI treatment, and washing were repeated five times to build up the desired multilayer structure. Finally, the device was annealed at 85 °C for 10 min.

3. Results

In this study, APTES was employed as an interfacial modification layer on ITO interdigitated electrodes to enhance the interaction between PbS QDs and the electrode surface, thereby addressing the gap in interface engineering for high-performance PbS QD-based SWIR photodetectors. As illustrated in Figure 1a, an APTES activation layer was introduced at the interface between the PbS QDs and the ITO electrodes. This modification aims to systematically investigate how changes at the interface affect the overall performance of the photodetector.

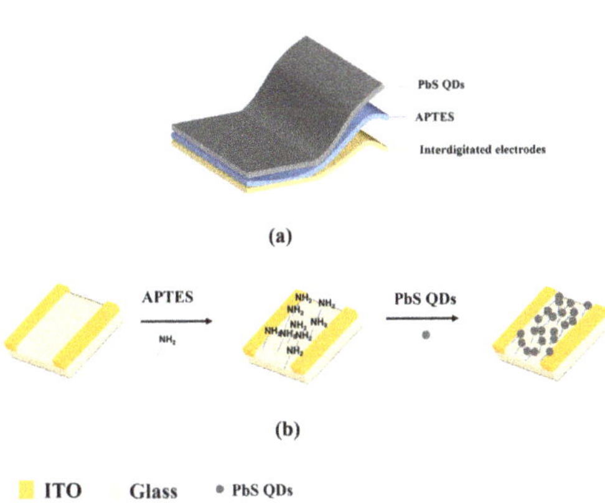

Figure 1. (**a**) Schematic structure of the SWIR photodetector device on ITO interdigitated electrode substrate; (**b**) illustration of the APTES optimization process.

Figure 1b presents a schematic diagram of the APTES-modified electrodes, highlighting the uniform coverage of the silane layer on the electrode surface. The modification process, detailed in Figure 1b, involves the covalent binding of APTES molecules onto the ITO surface via hydrolysis and condensation reactions, resulting in the formation of siloxane (Si-O-Si) bonds. This treatment exposes surface amino groups (-NH_2), which are given the chance to influence the surface chemistry and physical properties of the ITO electrodes. The presence of these amino groups provides an opportunity to enhance adhesion and promote a more uniform deposition of the PbS QD film compared to unmodified electrodes, offering a promising strategy to achieve a denser and more compact QD layer. This improvement in the QD film plays a vital role in promoting efficient charge transport while reducing carrier recombination losses, ultimately contributing to the enhanced performance of optoelectronic devices.

To validate this hypothesis, PbS QDs with a first excitonic absorption peak at 1270 nm were selected as the model system due to their well-defined optical properties in the short-wave infrared (SWIR) region. These QDs were employed as the photosensitive layer of the photodetector to investigate the impact of interfacial engineering on device performance. Figure 2 presents the TEM image and absorption spectrum of these QDs. As shown in Figure 2a,b, the PbS QDs are uniformly dispersed with an average diameter of 3.6 nm. Figure 2c shows the absorption spectrum of the QDs, featuring a sharp and well-defined first excitonic peak centered at 1270 nm. This peak is indicative of strong quantum confinement effects [30,31], which arise when the QD size approaches or falls below the exciton Bohr radius of bulk PbS. The distinct absorption behavior confirms that the QDs possess a bandgap well-suited for SWIR photodetection, making them promising candidates for high-performance optoelectronic applications such as photodetectors and image sensors.

We systematically investigated the impact of APTES introduction on the performance of PbS QD-based photodetectors. As illustrated in Figure 3, the ITO interdigitated electrodes were immersed in the APTES precursor solution for 0.5 h, and the resulting devices were compared with untreated counterparts. A comparison of the I–V characteristics (Figure 3a) reveals that the introduction of APTES creates a more favorable interfacial potential gradient, promoting more efficient separation and transport of photogenerated electron–hole pairs. This improvement is attributed to more efficient separation of photogenerated carriers, likely resulting from the formation of a denser and more uniform PbS QD film on the modified electrode surface. Furthermore, the responsivity–voltage (R–V) curves (Figure 3b) show a consistent increase in responsivity across the measured voltage range, further confirming the positive effect of the APTES layer in facilitating carrier transport. The on/off ratio (Figure 3c) analysis reveals that the LBL device exhibits a rise time of 14.98 ms and a fall time of 27.49 ms, while the APTES-treated device shows a rise time of 12.89 ms and a fall time of 24.25 ms. The results reflect that devices with an APTES duration of 0.5 h have a stronger response and faster response speed. This conclusion is further supported by the SEM analysis shown in Figure 4. In the unmodified device (Figure 4a,b), the PbS QD film exhibits a rough and granular surface morphology, characterized by irregular protrusions and visible voids. In contrast, the APTES-modified sample (Figure 4c,d) presents a significantly smoother and denser morphology. The grain boundaries become less distinct, and the porosity is markedly reduced. Concurrently, AFM analysis provides further evidence: (a) the unmodified device (Figure 5a) exhibits a higher friction force on the PbS QD film, while (b) the APTES-modified device (Figure 5b) shows reduced surface friction. This indicates that the smoothed surface post-modification suppresses carrier scattering, improves interfacial contact, and ultimately enhances device performance. These morphological improvements are indicative of enhanced film quality, which is critical for efficient charge transport in photodetector applications.

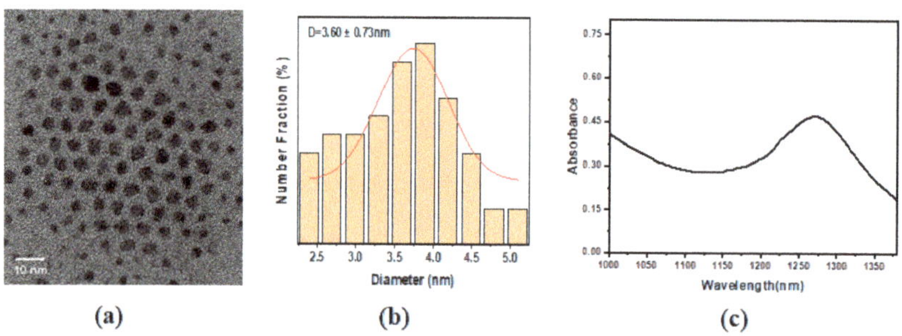

Figure 2. (**a**) TEM image of PbS QDs, scale bar: 10 nm; (**b**) size distribution histogram from TEM, average diameter: 3.6 nm; (**c**) absorption spectrum of PbS QDs.

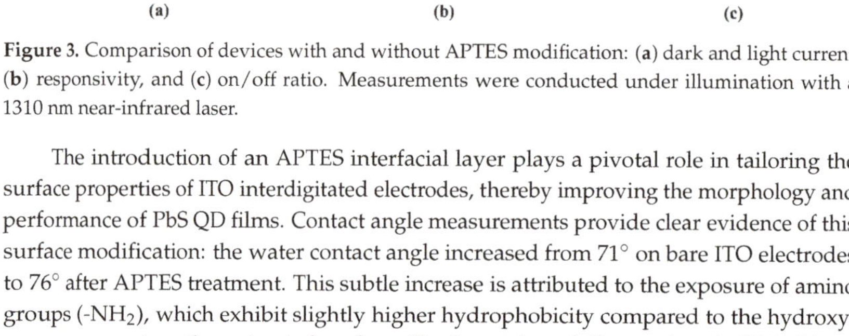

Figure 3. Comparison of devices with and without APTES modification: (**a**) dark and light current; (**b**) responsivity, and (**c**) on/off ratio. Measurements were conducted under illumination with a 1310 nm near-infrared laser.

The introduction of an APTES interfacial layer plays a pivotal role in tailoring the surface properties of ITO interdigitated electrodes, thereby improving the morphology and performance of PbS QD films. Contact angle measurements provide clear evidence of this surface modification: the water contact angle increased from 71° on bare ITO electrodes to 76° after APTES treatment. This subtle increase is attributed to the exposure of amino groups (-NH_2), which exhibit slightly higher hydrophobicity compared to the hydroxyl groups present on the untreated surface. The formation of siloxane bonds (Si-O-Si) via hydrolysis and condensation during the APTES modification process leads to the stable integration of these functional groups (Figure 1b). This change in surface wettability significantly affects the deposition behavior of the PbS QD film. Since these QDs are dispersed in a hydrophobic solvent such as n-octane and are capped with long-chain organic ligands (e.g., OA and OlAm), the enhanced hydrophobicity of the APTES-modified surface improves compatibility and adhesion between the QDs and the substrate. According to the principle of "like dissolves like", this improved surface match promotes the formation of a more uniform and compact film with reduced porosity and fewer defects. Figure 6a–c systematically investigate the effect of APTES modification time on the performance of PbS QDs photodetectors. By comparing devices fabricated under different soaking durations (0 h, 0.25 h, 0.5 h, 1 h, and 1.5 h), the current–voltage (I–V) characteristics under dark and 1310 nm monochromatic laser illumination, as well as the responsivity–voltage (R–V) curves, reveal the modulation mechanism of device performance by APTES interfacial modification.

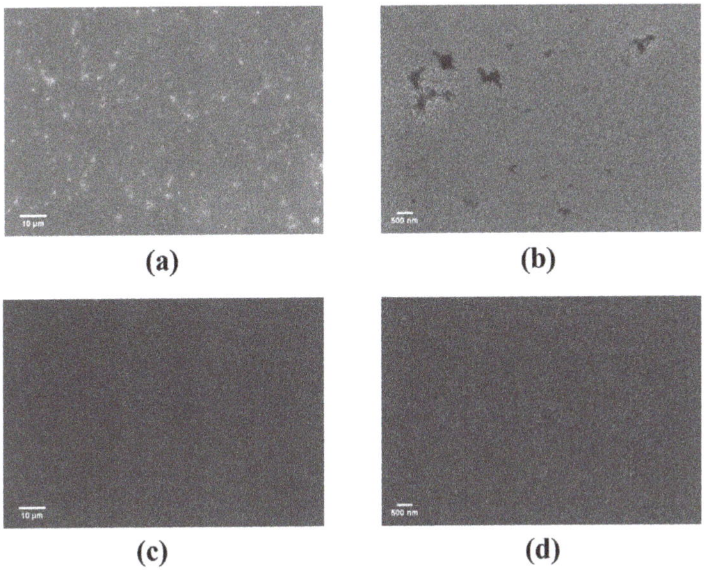

Figure 4. Comparison of SEM images: (**a**,**b**) unmodified devices vs. (**c**,**d**) APTES-modified devices. Scale bars: 10 μm (**a**,**c**); 500 nm (**b**,**d**).

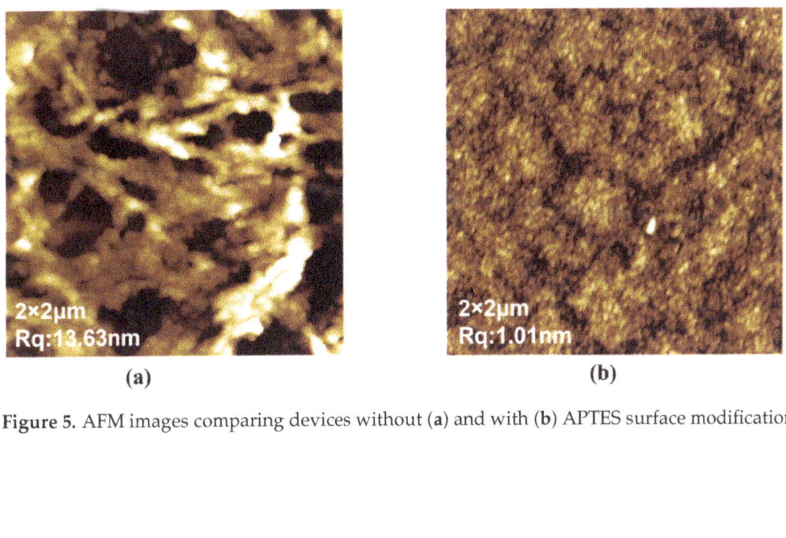

Figure 5. AFM images comparing devices without (**a**) and with (**b**) APTES surface modification.

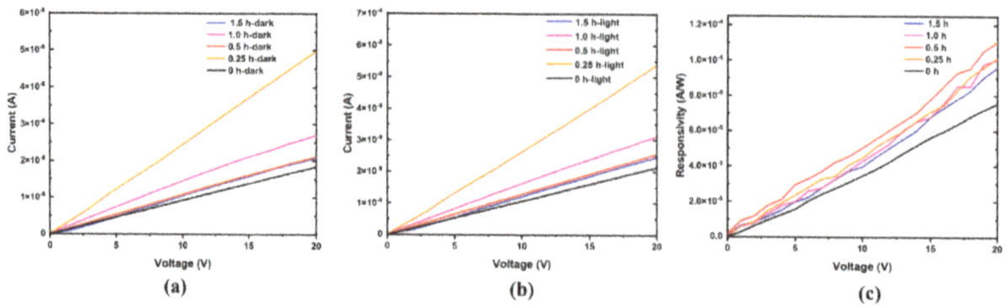

Figure 6. Comparison of device performance with different APTES modification durations: (**a**) dark current; (**b**) light current; (**c**) responsivity. Measurements were conducted under illumination with a 1310 nm near-infrared laser.

The results show that device performance exhibits a "rise then fall" trend as the APTES treatment time increases. At the initial stage, when the soaking time extends from 0 h to 0.25 h, the amino group density on the electrode surface increases, effectively passivating some interface defects, resulting in a significant reduction in dark current. Further extending the soaking time to 0.5 h leads to the formation of a dense and continuous silane layer, which improves the compactness and uniformity of the PbS QD film as well as the interface contact with the electrode. This greatly enhances the separation and collection efficiency of photogenerated carriers. During this stage, the device light current is significantly enhanced, and the responsivity correspondingly increases, with the 0.5 h treated sample exhibiting the best photo-response and overall device performance (Table 1). When combined with the detectivity (D*) results [19], the 0.5 h APTES treatment effectively enhances the D* improvement, strongly supporting the synergistic contributions of effective interfacial defect passivation and enhanced film densification. However, when the treatment time is further extended to 1 h and 1.5 h, the excessively thick silane layer may introduce additional tunneling barriers and increase the device's series resistance, thereby inhibiting carrier injection and transport. As a result, both light current and responsivity decline to varying degrees. This indicates that the effectiveness of APTES modification is highly sensitive to treatment duration; moderate modification optimizes interface quality and improves device performance, while excessive modification may cause adverse effects, hindering further enhancement. Therefore, precise control of APTES treatment time is crucial for achieving high-performance PbS QD photodetectors. These findings underscore the critical role of interface engineering in optimizing the performance of quantum dot-based optoelectronic devices.

Table 1. Comparison of device performance with different APTES modification durations [1].

Time (h)	Dark Current (A)	Light Current (A)	Responsivity(A/W)
0	1.83×10^{-8}	2.13×10^{-8}	7.54×10^{-5}
0.25	4.97×10^{-8}	5.37×10^{-8}	1.01×10^{-4}
0.5	2.11×10^{-8}	2.55×10^{-8}	1.10×10^{-4}
1.0	2.70×10^{-8}	3.10×10^{-8}	1.00×10^{-4}
1.5	2.07×10^{-8}	5.91×10^{-8}	9.57×10^{-5}

[1] At 20V and 1310 nm, the light intensity measured by the power meter is 0.834 mW; the working area of the optical power meter is 94.09 mm^2; and the optical power density is 0.89 mW/cm^2. The area of the photodetector is 4.5 mm^2 [32].

4. Conclusions

In summary, we propose an effective strategy for enhancing the performance of SWIR photodetectors through the APTES surface modification of ITO interdigitated electrodes. This modification significantly improves the interface between the QD film and the electrode, resulting in a denser and smoother film morphology. Such improvements are essential for promoting efficient carrier transport, minimizing recombination losses, and enabling stable, high-performance device operation. Notably, the APTES treatment time is a key parameter in the simultaneous optimization of current and responsivity. Insufficient modification leads to incomplete defect passivation, while excessive treatment introduces additional tunneling barriers and increases series resistance due to a thicker organic layer—both of which degrade device performance [33–35]. A 0.5 h treatment strikes the optimal balance, maximizing interfacial passivation and carrier transport. This optimized modification window provides a valuable reference for future device engineering. Further performance improvements may be achieved by combining interfacial energy level alignment or hybrid organic–inorganic dual-modification strategies, unlocking the full potential of PbS QD-based SWIR photodetectors for high-sensitivity, low-noise optoelectronic applications.

Author Contributions: Conceptualization, supervision, project administration, funding acquisition, J.H., J.C. and H.T.; methodology, investigation, Q.L., X.A. and F.F.; formal analysis, data curation, L.R. and W.D., writing—original draft preparation, Q.L. and H.T.; writing—review and editing, J.H. and W.C. All authors have read and agreed to the published version of the manuscript.

Funding: This research was funded by the National Natural Science Foundation of China (No. 62204107), Shenzhen Science and Technology Program (No. JCYJ20240813113240052, JCYJ20241202124709012), Project of Engineering Technology Center of General Universities of Guangdong province (No. 2023GCZX005), Pingshan District Innovation Platform Project of Shenzhen Hi-tech Zone Development Special Plan in 2022 (No. 29853M-KCJ-2023-002-01), and Natural Science Foundation of Top Talent of SZTU (grant no. GDRC202344, GDRC202345).

Data Availability Statement: The original contributions presented in the study are included in the article; further inquiries can be directed to the corresponding authors.

Acknowledgments: The author is grateful to the staff of the Analysis and Testing Center of Shenzhen Technology University for the TEM and absorption spectroscopy analysis.

Conflicts of Interest: The authors declare no conflicts of interest.

References

1. Moreels, I.; Justo, Y.; De Geyter, B.; Haustraete, K.; Martins, J.C.; Hens, Z. Size-Tunable, Bright, and Stable PbS Quantum Dots: A Surface Chemistry Study. *ACS Nano* **2011**, *5*, 2004–2012. [CrossRef] [PubMed]
2. Weidman, M.C.; Beck, M.E.; Hoffman, R.S.; Prins, F.; Tisdale, W.A. Monodisperse, Air-Stable PbS Nanocrystals via Precursor Stoichiometry Control. *ACS Nano* **2014**, *8*, 6363–6371. [CrossRef] [PubMed]
3. Wang, Z.; Gu, Y.; Li, X.; Liu, Y.; Liu, F.; Wu, W. Recent Progress of Quantum Dot Infrared Photodetectors. *Adv. Opt. Mater.* **2023**, *11*, 2300970. [CrossRef]
4. Zaini, M.S.; Liew, J.Y.C.; Alang Ahmad, S.A.; Mohmad, A.R.; Ahmad Kamarudin, M. Photoluminescence investigation of carrier localization in colloidal PbS and PbS/MnS quantum dots. *ACS Omega* **2020**, *5*, 30956–30962. [CrossRef] [PubMed]
5. Venettacci, C.; Martín-García, B.; Prato, M.; Moreels, I.; De Iacovo, A. Increasing responsivity and air stability of PbS colloidal quantum dot photoconductors with iodine surface ligands. *Nanotechnology* **2019**, *30*, 405204. [CrossRef] [PubMed]
6. Shrestha, A.; Batmunkh, M.; Tricoli, A.; Qiao, S.Z.; Dai, S. Near-Infrared Active Lead Chalcogenide Quantum Dots: Preparation, Post-Synthesis Ligand Exchange, and Applications in Solar Cells. *Angew. Chem. Int. Ed.* **2019**, *58*, 5202–5224. [CrossRef] [PubMed]
7. Pradhan, S.; Di Stasio, F.; Bi, Y.; Gupta, S.; Christodoulou, S.; Stavrinadis, A.; Konstantatos, G. High-efficiency colloidal quantum dot infrared light-emitting diodes via engineering at the supra-nanocrystalline level. *Nat. Nanotechnol.* **2019**, *14*, 72–79. [CrossRef] [PubMed]
8. Yang, X.; Yang, J.; Khan, J.; Deng, H.; Yuan, S.; Zhang, J.; Xia, Y.; Deng, F.; Zhou, X.; Umar, F.; et al. Hydroiodic Acid Additive Enhanced the Performance and Stability of PbS-QDs Solar Cells via Suppressing Hydroxyl Ligand. *Nano-Micro Lett.* **2020**, *12*, 37. [CrossRef] [PubMed]
9. Tavakoli Dastjerdi, H.; Tavakoli, R.; Yadav, P.; Prochowicz, D.; Saliba, M.; Tavakoli, M.M. Oxygen plasma-induced p-type doping improves performance and stability of PbS quantum dot solar cells. *ACS Appl. Mater. Interfaces* **2019**, *11*, 26047–26052. [CrossRef] [PubMed]

10. Chen, W.; Guo, R.; Tang, H.; Wienhold, K.S.; Li, N.; Jiang, Z.; Tang, J.; Jiang, X.; Kreuzer, L.P.; Liu, H.; et al. Operando structure degradation study of PbS quantum dot solar cells. *Energy Environ. Sci.* **2021**, *14*, 3420–3429. [CrossRef]
11. Imamura, Y.; Yamada, S.; Tsuboi, S.; Nakane, Y.; Tsukasaki, Y.; Komatsuzaki, A.; Jin, T. Near-infrared emitting PbS quantum dots for in vivo fluorescence imaging of the thrombotic state in septic mouse brain. *Molecules* **2016**, *21*, 1080. [CrossRef] [PubMed]
12. Benayas, A.; Ren, F.; Carrasco, E.; Marzal, V.; del Rosal, B.; Gonfa, B.A.; Juarranz, Á.; Sanz-Rodríguez, F.; Jaque, D.; García-Solé, J.; et al. PbS/CdS/ZnS Quantum Dots: A Multifunctional Platform for In Vivo Near-Infrared Low-Dose Fluorescence Imaging. *Adv. Funct. Mater.* **2015**, *25*, 6650–6659. [CrossRef]
13. Zebibula, A.; Alifu, N.; Xia, L.; Sun, C.; Yu, X.; Xue, D.; Liu, L.; Li, G.; Qian, J. Ultrastable and Biocompatible NIR-II Quantum Dots for Functional Bioimaging. *Adv. Funct. Mater.* **2018**, *28*, 1703451. [CrossRef]
14. Pejović, V.; Georgitzikis, E.; Lee, J.; Lieberman, I.; Cheyns, D.; Heremans, P.; Malinowski, P.E. Infrared Colloidal Quantum Dot Image Sensors. *IEEE Trans. Electron Devices* **2022**, *69*, 2840–2850. [CrossRef]
15. Georgitzikis, E.; Malinowski, P.E.; Li, Y.; Maes, J.; Hagelsieb, L.M.; Guerrieri, S.; Hens, Z.; Heremans, P.; Cheyns, D. Integration of PbS Quantum Dot Photodiodes on Silicon for NIR Imaging. *IEEE Sens. J.* **2020**, *20*, 6841–6848. [CrossRef]
16. Baranov, D.; Lynch, M.J.; Curtis, A.C.; Carollo, A.R.; Douglass, C.R.; Mateo-Tejada, A.M.; Jonas, D.M. Purification of Oleylamine for Materials Synthesis and Spectroscopic Diagnostics for trans Isomers. *Chem. Mater.* **2019**, *31*, 1223–1230. [CrossRef]
17. Chen, S.; Zhong, H.; Wang, X.; Pan, G.; Tang, H.; Fang, F.; Wu, J.; Wang, W.; Xu, L.; Tang, J.; et al. Hybrid-Size Quantum Dots in Hole Transport Layer Depress Dark Current Density of Short-Wave Infrared Photodetectors. *ACS Photonics* **2025**, *12*, 879–888. [CrossRef]
18. Tang, H.; Zhong, J.; Chen, W.; Shi, K.; Mei, G.; Zhang, Y.; Wen, Z.; Müller-Buschbaum, P.; Wu, D.; Wang, K.; et al. Lead Sulfide Quantum Dot Photodetector with Enhanced Responsivity through a Two-Step Ligand-Exchange Method. *ACS Appl. Nano Mater.* **2019**, *2*, 6135–6143. [CrossRef]
19. Zhang, W.; Fang, F.; Zhong, H.; Huang, L.; Tang, H.; Chen, X.; Hao, J.; Zhang, L.; Cao, L.; Tang, J.; et al. Converting Perovskite Nanocrystals to PbS Quantum Dots Toward Short-Wave Infrared Photodetectors. *Adv. Opt. Mater.* **2025**, *13*, 2402740. [CrossRef]
20. Chen, C.; Ao, L.; Wu, Y.T.; Cifliku, V.; Cardoso Dos Santos, M.; Bourrier, E.; Delbianco, M.; Parker, D.; Zwier, J.M.; Huang, L.; et al. Single-Nanoparticle Cell Barcoding by Tunable FRET from Lanthanides to Quantum Dots. *Angew. Chem. Int. Ed.* **2018**, *57*, 13686–13690. [CrossRef] [PubMed]
21. Pietra, F.; van Dijk-Moes, R.J.A.; Ke, X.; Bals, S.; Van Tendeloo, G.; de Mello Donega, C.; Vanmaekelbergh, D. Synthesis of Highly Luminescent Silica-Coated CdSe/CdS Nanorods. *Chem. Mater.* **2013**, *25*, 3427–3434. [CrossRef]
22. Koole, R.; Van Schooneveld, M.M.; Hilhorst, J.; de Mello Donegá, C.; Hart, D.C.; Van Blaaderen, A.; Vanmaekelbergh, D.; Meijerink, A. On the incorporation mechanism of hydrophobic quantum dots in silica spheres by a reverse microemulsion method. *Chem. Mater.* **2008**, *20*, 2503–2512. [CrossRef]
23. Huang, P.Y.; Zhang, Y.Y.; Tsai, P.C.; Chung, R.J.; Tsai, Y.T.; Leung, M.K.; Lin, S.Y.; Fang, M.H. Interfacial Engineering of Quantum Dots–Metal–Organic Framework Composite Toward Efficient Charge Transport for a Short-Wave Infrared Photodetector. *Adv. Opt. Mater.* **2024**, *12*, 2302062. [CrossRef]
24. Wu, Z.; Ou, Y.; Cai, M.; Wang, Y.; Tang, R.; Xia, Y. Short-wave infrared photodetectors and imaging sensors based on lead chalcogenide colloidal quantum dots. *Adv. Opt. Mater.* **2023**, *11*, 2201577. [CrossRef]
25. Xu, K.; Zhou, W.; Ning, Z. Integrated structure and device engineering for high performance and scalable quantum dot infrared photodetectors. *Small* **2020**, *16*, 2003397. [CrossRef] [PubMed]
26. Yin, X.; Zhang, C.; Guo, Y.; Yang, Y.; Xing, Y.; Que, W. PbS QD-based photodetectors: Future-oriented near-infrared detection technology. *J. Mater. Chem. C* **2021**, *9*, 417–438. [CrossRef]
27. Yuan, Y.; Xu, J.-L.; Zhang, J.-Y.; Gao, X.; Zhong, Y.-N.; Wang, S.-D. Interface engineering for high photoresponse in PbS quantum-dot short-wavelength infrared photodiodes. *IEEE Electron Device Lett.* **2022**, *43*, 1275–1278. [CrossRef]
28. Hines, M.A.; Scholes, G.D. Colloidal PbS Nanocrystals with Size-Tunable Near-Infrared Emission: Observation of Post-Synthesis Self-Narrowing of the Particle Size Distribution. *Adv. Mater.* **2003**, *15*, 1844–1849. [CrossRef]
29. Cao; Banin, U. Growth and properties of semiconductor core/shell nanocrystals with InAs cores. *J. Am. Chem. Soc.* **2000**, *122*, 9692–9702. [CrossRef]
30. Luther, J.M.; Zheng, H.; Sadtler, B.; Alivisatos, A.P. Synthesis of PbS nanorods and other ionic nanocrystals of complex morphology by sequential cation exchange reactions. *J. Am. Chem. Soc.* **2009**, *131*, 16851–16857. [CrossRef] [PubMed]
31. Reiss, P.; Protiere, M.; Li, L. Core/shell semiconductor nanocrystals. *Small* **2009**, *5*, 154–168. [CrossRef] [PubMed]
32. Saran, R.; Curry, R.J. Lead sulphide nanocrystal photodetector technologies. *Nat. Photonics* **2016**, *10*, 81–92. [CrossRef]
33. Saengdee, P.; Promptmas, C.; Thanapitak, S.; Srisuwan, A.; Pankiew, A.; Thornyanadacha, N.; Chaisriratanakul, W.; Chaowicharat, E.; Jeamsaksiri, W. Optimization of 3-aminopropyltriethoxysilane functionalization on silicon nitride surface for biomolecule immobilization. *Talanta* **2020**, *207*, 120305. [CrossRef] [PubMed]

34. Kim, J.; Seidler, P.; Wan, L.S.; Fill, C. Formation, structure, and reactivity of amino-terminated organic films on silicon substrates. *J. Colloid Interface Sci.* **2009**, *329*, 114–119. [CrossRef] [PubMed]
35. Howarter, J.A.; Youngblood, J.P. Optimization of Silica Silanization by 3-Aminopropyltriethoxysilane. *Langmuir* **2006**, *22*, 11142–11147. [CrossRef] [PubMed]

colloids and interfaces

MDPI

Article

Rapid Removal of Sizing Agent from Carbon Fiber Surface by Liquid-Phase Plasma Electrolysis

Chiyuhao Huang [1], Qian Zhou [1], Maoyuan Li [2], Xiaolin Wei [1], Dongqin Li [1], Xin He [1] and Weiwei Chen [1,*]

[1] Department of Materials Science and Engineering, Beijing Institute of Technology, Beijing 100081, China; cyh_huang@163.com (C.H.); 17812162775@163.com (Q.Z.); 3120235757@bit.edu.cn (X.W.); 15210581440@163.com (D.L.); xin_he0205@163.com (X.H.)

[2] Beijing System Design Institute of Electro-Mechanic Engineering, Beijing 100854, China; maoyuan_li@163.com

* Correspondence: wwchen@bit.edu.cn

Abstract

In this study, liquid-phase plasma electrolysis (LPE) was successfully employed to remove the sizing agent from T800 carbon fiber surfaces. Through systematic investigation of varying arcing voltages (185–215 V) and electrode spacings (10–20 mm), we determined that an optimal combination of 200 V and 10 mm spacing achieved near-complete sizing removal, as confirmed by SEM, TGA, and XPS analyses. Under this condition, plasma bombardment dominated the removal mechanism, eliminating sizing residues while exposing the underlying fiber grooves. TGA further demonstrated that in samples treated at a 10 mm interval, the weight loss of LPE samples before 300 °C was negligible, indicating that the sizing agent had been thoroughly removed. The results of XPS further confirmed the high efficiency of LPE in the removal of sizing agents (C-O bond content from 41.6% to 26.9%), and the retention of C-O also proved that LPE could maintain the surface activity of carbon fibers, confirming the effectiveness of LPE in decomposing the sizing agent. Meanwhile, based on the above test results, an attempt was made to explain the mechanism of LPE in removing sizing agents from the surface of carbon fibers.

Keywords: plasma; carbon fiber; sizing agent removal

1. Introduction

Carbon fiber reinforced composite materials (CFRP) feature high strength, good thermal conductivity, and high-temperature dimensional stability, which can meet the demands of the aerospace field [1–4]. They are often used in parts such as the head cone, leading edge, and throat liner of high-speed aircraft [5]. Meanwhile, its ease of modification, high adjustability, and anisotropy in performance make it have broad application prospects in energy and multi-functional surfaces [6–8]. PAN-based carbon fibers, as the most representative product among carbon fibers [9], are commonly used as reinforcing materials in CFRP. During the production process of carbon fibers, oxidation treatment is carried out to introduce oxygen-containing groups on the inert graphite surface of the carbon fibers, facilitating the subsequent processing of composite materials [10–12]. Then, it still needs to go through the sizing step to ensure that the carbon fibers do not experience partial fiber breakage and pulping or fiber dispersion during transportation, and at the same time, it may be beneficial for the subsequent processing of composite materials [13–16]. However, in the specific processing, in order to make the surface state of the carbon fiber more in line with the corresponding processing technology [17], or to increase the degree of graphitization on the surface of the carbon fiber [18], or to increase the surface roughness of the carbon fiber [19], the sizing agent on the surface of the carbon fiber needs to be removed. Fang et al. [20] removed the sizing agent on the surface of carbon fibers by soaking them in anhydrous ethanol for 7 h, ensuring that the glucose on the surface of the carbon fibers was subsequently converted into carbon. Niranjan Patra et al. [21] used acetone plus ultrasonic treatment to remove the sizing agent on the surface of carbon fiber fabric, thereby ensuring the quality of the subsequent chemically deposited Ni-Co-Fe-P coating. The common method for removing sizing agents in production is to subject carbon fibers to heat treatment at high temperatures [22], in air or an inert atmosphere, for a period of time. This method is limited by the volume of the heating container, and production is difficult to be carried out continuously. Moreover, high temperatures can easily damage the active groups on the surface of carbon fibers, weaken the interfacial adhesion, and reduce the interfacial strength between carbon fibers and CFRP, thereby affecting the fracture behavior of the composite material [23,24]. Plasma surface modification of carbon fibers can introduce oxygen-containing groups and increase the surface activity of carbon fibers [25–29]. Lee et al. [30] used hydrogen plasma to modify the surface of carbon fibers, enhancing the interfacial adhesion of CFRP. Xiao et al. [31] developed a continuous and rapid atmospheric plasma system with power and successfully introduced oxygen-containing functional groups on the CF surface. Moosburg-Will et al. [32] successfully and continuously modified carbon fibers by using the atmospheric plasma jet method (APPJ). However, some of these methods have high requirements for the airtightness of high-pressure gas containers and too many parameters that need to be regulated, which limits their application in production.

In this paper, liquid-phase plasma electrolysis is used to remove the sizing agent from the surface of carbon fibers. This not only meets the conditions for continuous production but also avoids the risks of high-pressure gas containers. Meanwhile, the liquid environment can reduce the damage to the fiber surface, and the free groups in the liquid can enhance the surface activity of carbon fibers. The existence form and distribution of the sizing agent on the surface of carbon fibers were observed by scanning electron microscopy (SEM). The content of sizing agent on the surface of each carbon fiber sample was characterized by a thermogravimetric analysis test (TG). The types and relative contents of carbon-containing chemical bonds on the surface of carbon fibers were characterized by X-ray photoelectron spectroscopy (XPS). The optimal process for removing the sizing agent from the surface of carbon fibers with LPE was obtained. It was confirmed that LPE can completely remove the sizing agent from the surface of carbon fibers while introducing oxygen-containing groups. It provides more effective methods and means for improving the interface performance and production efficiency of CFRP.

2. Experimental Section

The experimental procedures of this study are illustrated in Figure 1. Carbon fibers with sizing agent (T800, HF40S-12K, Jiangsu Hengshen Co., Ltd., Zhenjiang, China) in their bare form were subjected to liquid-phase plasma electrolysis (LPE) treatment under varying process parameters to obtain samples a–i (Table 1). Subsequently, the surface sizing agent content of both the bare fibers and samples a–i were characterized using scanning electron microscopy (SEM), thermogravimetric analysis (TG), and X-ray photoelectron spectroscopy (XPS).

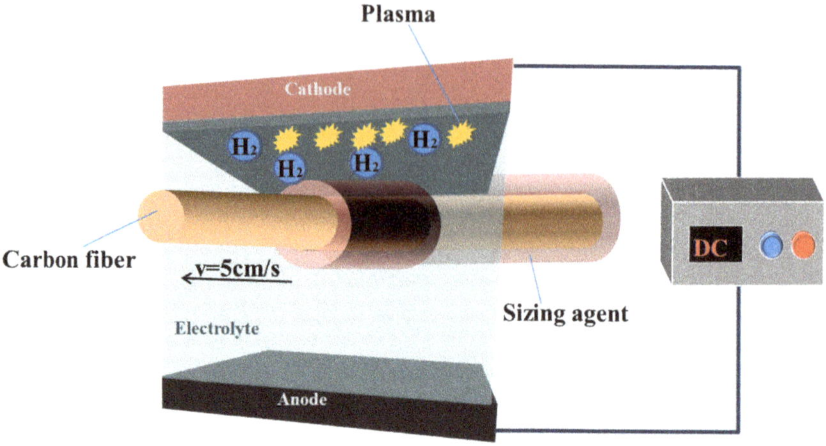

Figure 1. Schematic diagram of the process of removing sizing agent from the surface of carbon fibers by LPE.

Table 1. Labels of samples to their parameters.

Electrode Spacing \ Arcing Voltage	185 V	200 V	215 V
20 mm	a	b	c
15 mm	d	e	f
10 mm	g	h	i

2.1. LPE Process

The LPE apparatus employed in this study is illustrated in Figure 2. The system is equipped with a TN-KGZ01 high-frequency switching DC power supply (Guochong Charging Technology Co., Ltd., Yangzhou, China) capable of providing continuously adjustable voltage output ranging from 0 to 1500 V. A stepper motor is incorporated to ensure continuous speed regulation of the carbon fibers between 2.8 and 20 cm/s. The cathode rod assembly features adjustable studs at both ends, enabling precise control of the electrode-to-fiber distance within a range of 10–20 mm.

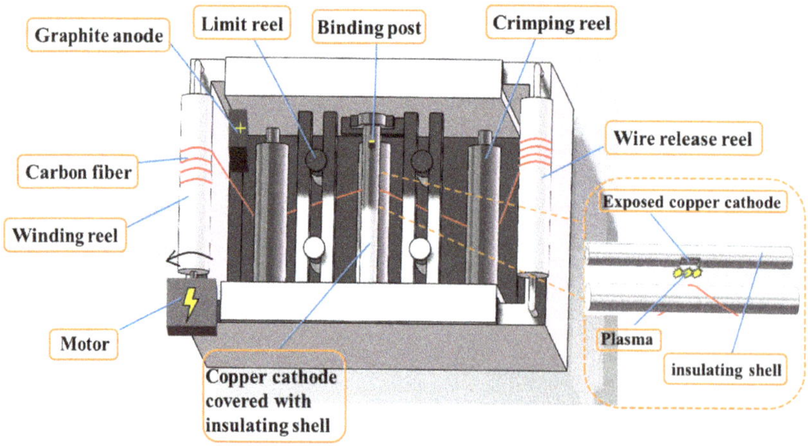

Figure 2. Equipment structure diagram.

During LPE treatment, a sodium metasilicate solution (50 g/L) serves as the liquid medium to establish an electric field between the cathode and graphite anode, thereby generating plasma beneath the cathode, as shown in Figure 3. The use of 50 g/L sodium metasilicate as the liquid medium is mainly because sodium metasilicate has a high solubility in water, which can provide sufficient cations and anions for electron exchange at the electrode in the solution. Moreover, no by-products will be formed during the electrolysis process, ensuring the continuity of plasma generation. The carbon fibers, maintained at a constant speed of 5 cm/s, are precisely guided through the plasma zone directly below the cathode via the combined action of tensioning rollers and positioning guides. During the experiment, the electric field force caused the free hydrogen ions in the water to approach the cathode, continuously generating hydrogen bubbles near the cathode. When the voltage reaches the breakdown voltage of the hydrogen bubble, the hydrogen bubble breaks down, forming plasma. Because carbon fibers also have electrical conductivity, hydrogen bubbles are precipitated from the surface of the carbon fibers when they are close to the cathode. These hydrogen bubbles will also turn into plasma due to high voltage. Therefore, plasma exists between the cathode and the carbon fiber. When the carbon fiber continuously passed through the solution, we considered that the plasma was continuously treating the surface of the carbon fiber.

Figure 3. Equipment: real figure at work.

2.2. Characterization

The surface morphology of carbon fibers, along with the morphology and the form of the sizing agent, was characterized using field-emission scanning electron microscopy (EHT = 15.00 K, WD = 9.4 mm, FE-SEM, GeminiSEM 300 ZEISS, Oberkohen, Germany).

Thermogravimetric analysis (TGA) was conducted using the thermal analysis system TGA/DSC 3 (air 20 mL/min, heating rate 10 k/min, temperature from 25 °C to 400 °C, METTLER Toledo, Zurich, Switzerland) in an air atmosphere to screen out the optimal LPE samples based on weight loss at 300 °C. Then, the selected samples were heated to 450 °C using the same equipment and kept at this temperature for 20 min to obtain the curve of their weight varying with time. Furthermore, X-ray photoelectron spectroscopy (XPS, QUANTERA-II SXM, PHI, Ulvac-Phi Inc., Chigasaki, Japan) was employed to analyze the relative contents of carbon chemical bonds and the C/O ratio on the surfaces of both sized and desized carbon fiber samples. The tensile strength and retention rate of LPE fibers were tested using a universal mechanical testing machine (Instron5966, Boston, MA, USA).

3. Results and Discussion

3.1. Morphology and Sizing Agent Content Estimation

The numbers 1 to 3 marked in the lower left corner of the picture correspond to the 500×, 2000×, and 10,000× scanning electron microscope images of the sample, respectively. HF40S filament is prepared by a wet spinning process. The spinning solution directly enters the coagulation bath. At this time, the elastic gel surface formed on the surface of the solidified filament will elongate along the axial direction and contract along the radial direction under the action of axial tensile force, causing the solidified filament to become fine and denier, and the cross-sectional area to continuously decrease, thus forming folds and wrinkles on the surface. Meanwhile, the solvent inside the solidified filaments is gradually replaced by the coagulant, namely water, which leads to collapse and increases the driving force for the formation of wrinkles. Grooves on the fiber surface are formed during the solidification process and are oriented and arranged along the fiber axis under the action of uniaxial stretching force, and eventually remain on the surface of the carbon fiber.

By comparing the bare sample with samples a–i, it can be seen that after LPE treatment, there are residues of varying degrees on the surface, and the grooves on the surface of the carbon fiber monofilament are exposed. In the 2000× photo of the bare sample, some fine filament fragments scattered on the surface of the fibers can be observed. These fragments might be due to the fact that the sizing agent solidified during the sizing process and did not penetrate into the gap between the two closely attached filaments. Instead, it solidified on the surface of the two closely attached filaments due to surface tension. After the long fiber bundles were cut, the sizing agent at the bonding area was scattered, maintaining a part of the straight shape. The formation of these patterns may be attributed to the residue of the sizing agent. By comparing the 10,000× photos of the exposed sample and a–i, it can be found that the monofilament surface of the exposed sample has no pattern along the length direction, and the edge of the image is not a straight line either, indicating that there is something wrapped around the cylindrical surface. We believe that this is the sizing agent. The carbon fiber monofilament surface of the a–i sample shows grooves along the length direction, and the diameter is relatively uniform. This is because there is a coating of sizing agent on the surface of bare, which fills the grooves and patterns on its surface and makes it appear relatively smooth. At the same time, due to the dispersion of carbon fiber monofilaments, the sizing agent between the monofilaments is difficult to be evenly separated, resulting in uneven diameters of the monofilaments. Additionally, after the LPE treatment of samples a–i, the sizing agent layer on the surface is removed to varying degrees. This exposes the grooves on the surface of the a–i carbon fiber monofilaments in the sample, and the diameters are relatively uniform.

Samples a–c are the samples with voltages of 185 V, 200 V, and 215 V, respectively, when they are 20 mm away from the electrode (Figure 4). By observing samples a to c, it can be seen that from sample a to sample c, the residue on the surface of the carbon fiber is getting less and less. This is because when the distance from the electrode is 20 mm, the plasma arc region cannot completely affect the carbon fiber. At this point, the main influences on the fibers are the temperature field of the electrode and the mechanical vibrations during the bubble generation process. Therefore, when the voltage rises, the temperature of the electrode increases accordingly, and the generation rate of bubbles accelerates, thereby speeding up the melting and decomposition of the sizing agent.

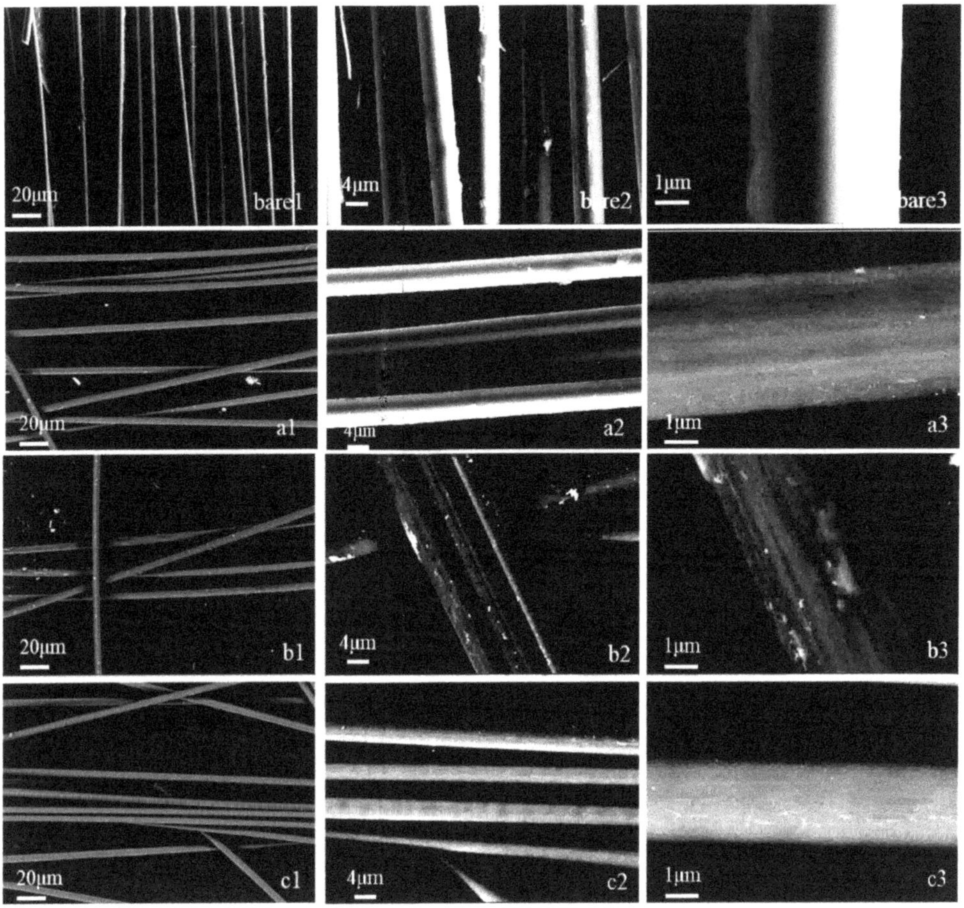

Figure 4. Scanning electron microscope images of the bare sample and samples a–c; 1, 2, and 3 are images at 500×, 2000×, and 10,000× respectively.

Samples d–f are the samples with voltages of 185 V, 200 V, and 215 V, respectively, when they are 15 mm away from the electrode (Figure 5). From photos d2–f2, point-like sizing agents with relatively high contrast on the surface of the fiber monofilaments can still be observed. When observing the fiber monofilaments at a magnification of 10,000×, it was found that the axial grooves on the fiber surface were relatively obvious, but there was still a small amount of residue remaining in the form of thin layers or spots. By comparing the surface morphology of samples d–f, it can be inferred that the change in the current voltage has no obvious effect on the surface morphology of carbon fibers. This might be because when the distance between the fiber and the electrode is 15 mm, the fiber is not only affected by the temperature field of the electrode and the mechanical vibration generated by the bubble, but also directly bombarded by the plasma arc. Under the influence of temperature, the surface sizing agent melts. Under the influence of mechanical vibration, the sizing agent layer splits. Under the direct bombardment of plasma, the sizing agent vaporizes and disintegrates. At this point, the influence of direct plasma bombardment begins to dominate. However, due to the relatively long distance between the fibers and the electrode, the effect of direct plasma bombardment remains weak, which results in a small amount of sizing agent remaining on the fiber surface.

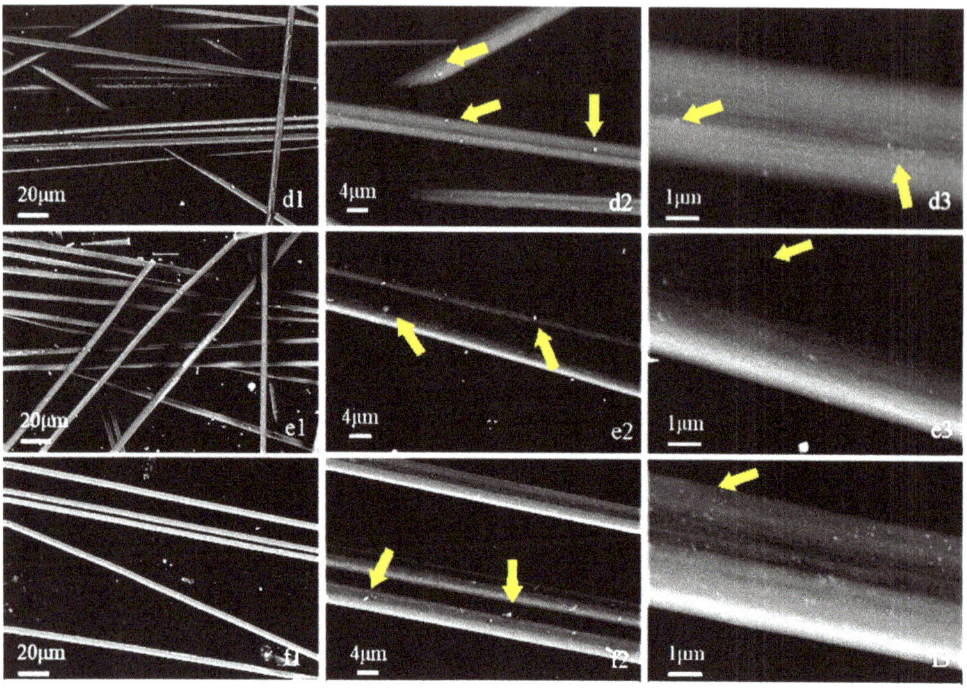

Figure 5. Scanning electron microscope images of samples d–f, 1, 2, and 3 are images at 500×, 2000×, and 10,000× respectively, some of the residues are marked by yellow arrows.

Samples g–i are the samples with voltages of 185 V, 200 V, and 215 V, respectively, when they are 15 mm away from the electrode (Figure 6). From the photos of samples g–i, we can see that from 500× to 10,000×, the surfaces of carbon fibers are all very smooth, and the shapes are all straight cylinders. At 10,000×, the grooves on the surface of the fibers are very clear. Although there are still some residues in samples g and i, they are much less than those in d–f. This indicates that reducing the distance from 15 mm to 10 mm is effective for the removal of the sizing agent. This is because when the fiber is 10 mm away from the electrode, the intensity and energy of the plasma are the greatest, and the sizing agent on the surface of the carbon fiber disintegrates and vaporizes after being directly bombarded by the plasma. By comparing samples a–i, it is concluded that for removing the sizing agent, 10 mm or 15 mm might be the more suitable distance between the carbon fiber and the electrode.

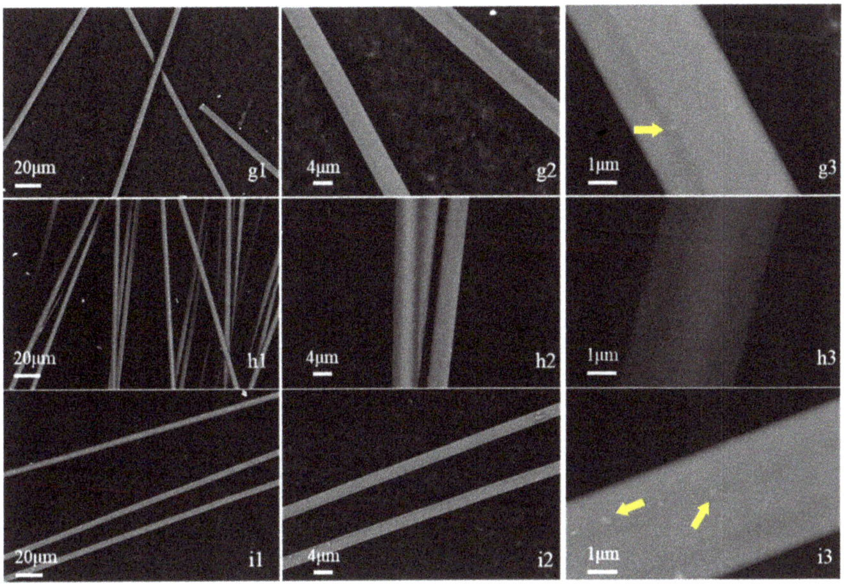

Figure 6. Scanning electron microscope images of samples g–i, 1, 2, and 3 are images at 500×, 2000×, and 10,000× respectively, some of the residues are marked by yellow arrows.

3.2. Thermogravimetric Analysis

The TGA curves of all samples were obtained under an air atmosphere with a heating rate of 10 °C/min up to 400 °C, as shown in Figure 7. Common commercial sizing agents may include one or more of PEEK, diazogen-containing diamine, dodecamine-modified epoxy resin, PUR, PA, PI, PHE, PVP, PES, E-MA-GMA, PPhEK, and NSM, which are dissolved in one or more of EP, CYC, THF, Toluene, DMC, and NMP; they may be in the form of an emulsion or an aqueous solution. It was found that when heated to 300 °C in the air, the volatile components of the sizing agent surrounding the carbon fiber completely evaporated.

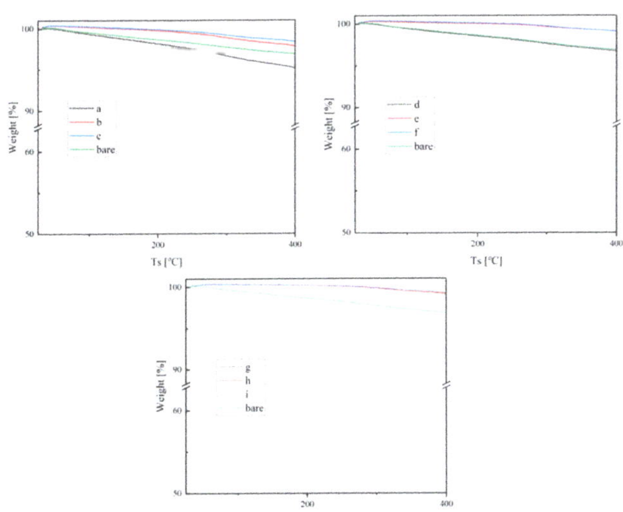

Figure 7. TG of the bare sample and samples a–i.

When the temperature rises from 25 °C to 300 °C, the evaporation of the sizing agent on the surface of carbon fibers is the main cause of weight loss. When the temperature exceeds 300 °C, the oxidation of carbon fibers becomes the main cause of weight loss. This is why the bare sample is constantly losing weight. Because there is a considerable amount of the sizing agent remaining on the surface of the bare sample, which causes it to show weight loss in both stages. Samples a–c are the samples with voltages of 185 V, 200 V, and 215 V, respectively, when they are 20 mm away from the electrode. It can be seen from the weight loss curves of samples a–c that when the spacing is 20 mm, as the voltage increases from 185 V to 215 V, the weight loss rate of the samples becomes smaller and smaller. This might be because when the spacing is 20 mm, the carbon fibers in LPE are mainly affected by the mechanical vibration caused by the electrode temperature field and bubble generation, while the effect of plasma bombardment is not significant.

Samples d–f are the samples with voltages of 185 V, 200 V, and 215 V, respectively, when they are 15 mm away from the electrode. By comparing the weight loss curves of samples d–f, it can be seen that the weight loss rate of sample d is significantly greater than that of samples e and f, which is very straight between 25 and 300 °C, indicating that the sizing agent of samples g–i has been removed completely, and the weight loss rates of samples e and f are similar. This evidence shows that when the spacing is 15 mm and the voltage is 185 V, the plasma still cannot fully affect the carbon fiber. Possibly, when the voltage increases to 200 V and 215 V, plasma bombardment begins to become the main factor affecting the removal efficiency of the sizing agent.

Samples g–i are the samples with voltages of 185 V, 200 V, and 215 V, respectively, when they are 10 mm away from the electrode. It can be seen from the weight loss curve of samples g–i that the curve of samples g–i is very straight between 25 and 300 °C, which indicates that the sizing agent of samples g–i has been removed completely. This indicates that when the distance between the carbon fiber and the electrode is 10 mm, the carbon fiber can be fully affected by the plasma. Maybe at this time, plasma bombardment is the main factor affecting the removal efficiency of the sizing agent. By comparing the weight loss curves of samples g–i and e and f, it can be seen that their weight loss curves have the same shape. It can be concluded that samples e–i are related to the weight loss curve of pure carbon fiber without the sizing agent. By comparing the weight loss curves of the bare sample, it can be seen that only the weight loss rate of sample a is greater than that of the bare sample, while the weight loss rates of the remaining samples are all less than that of the bare sample. This might be because under the conditions of a spacing of 20 mm and a voltage of 185 V, the plasma not only failed to bombard the sizing agent but also scratched the sizing agent layer, increasing the contact area with the air and causing the weight loss rate of sample a to be greater than that of the bare sample.

Since the weight loss curves of samples e–i are very similar, in order to obtain the optimal parameters for removing the sizing agent, it is necessary to compare the remaining weights of each sample at 300 °C (Figure 8). As shown in the figure, the remaining weight of each sample at 300 °C clearly indicates that the optimal parameter range for removing the sizing agent is an electrode spacing of 10 mm and an applied voltage of 185–200 V.

To determine the ability of the LPE method and the high-temperature method to remove the sizing agent, sample h, the bare sample, and CF that had undergone 1 h heat treatment in a 700 °C nitrogen atmosphere were tested. The temperature was increased from 30 °C to 450 °C at 10 K/min in a nitrogen atmosphere and then held at 450 °C for 20 min [33]. The curve of weight varying with time is shown in Figure 9.

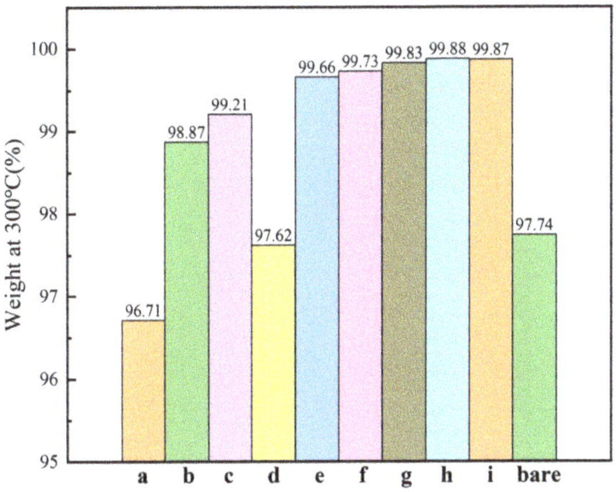

Figure 8. The remaining weight of each sample at 300 °C.

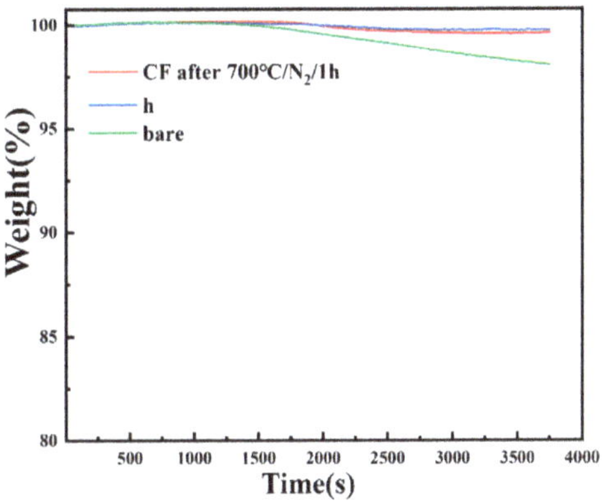

Figure 9. Weight-time curves of sample h, the bare sample, and CF after 700 °C/N_2/1 h.

As can be seen from Figure 9, during the heating stage 1 (0 < t < 1500 s), the weights of the three samples hardly changed, and the temperature at this time was 280 °C. During the heating stage 2 (1500 < t < 2520 s), the bare sample began to lose weight continuously, while the thermal decomposition method sample and sample h maintained their weight. During the holding stage (t > 2520 s), the bare sample continued to lose weight, and the weights of the thermal decomposition method sample and sample h remained almost unchanged. In a nitrogen atmosphere below 450 °C, the continuous weight loss of the bare sample is due to the decomposition and volatilization of the sizing agent, which indicates that the effect of LPE in removing the sizing agent is the same as that of the thermal decomposition method.

3.3. XPS

XPS analysis was performed on sample h, the bare sample, and CF after 700 °C/N_2/1 h (Al target, 1486.6 eV, line width < 0.48 eV), as shown in Figure 10 and Table 2. It can be seen from the XPS results that the carbon-containing chemical bonds on the surface of carbon fibers mainly exist in the form of C-C and C-O. However, on the CF surface where the sizing agent is removed at high temperature, there is no presence of C-O. The relative content ratio of C-C/C-O on the bare carbon fiber surface was 58.4/41.6. The relative content ratio of C-C/C-O on the surface of carbon fibers after the sizing agent was removed was 77.4/22.6. It can be seen that after LPE treatment, the content of C-O on the surface of carbon fibers decreased. This is because the C-O on the surface of carbon fibers mainly exists in the sizing agent. After LPE, the relative content of C-O decreased significantly as the sizing agent was completely removed. However, due to the etching effect of plasma on the surface of carbon fibers and the protection of the liquid, some oxygen-containing groups were retained on the surface of carbon fibers. Thus, some C-O were retained. This is conducive to the combination of carbon fibers and the subsequent production of composite materials.

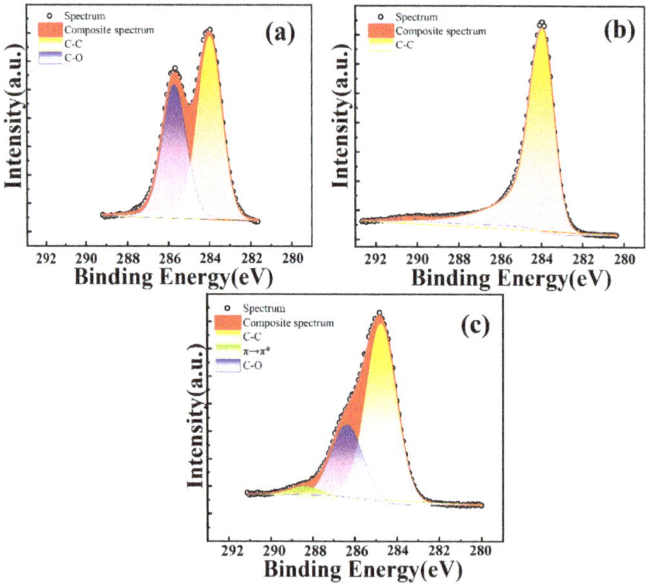

Figure 10. The peak fitting results: (**a**) bare sample; (**b**) CF after 700 °C/N_2/1 h; (**c**) sample h.

Table 2. The relevant peak surface parameters of the bare sample and sample h.

Sample	Binding Energy/eV	Peak Height	Full Width at Half Maximum (FWHM)	Peak Area	Percentage of Peak Area
h	284.0 (C-Csp2) 285.64 (C-O)	6490 2630	1.71 1.76	13,057 4811	77.4 22.6
CF after 700 °C /N_2/1 h	284.0 (C-Csp2)	13,928	1.28	25,678	100
bare	284.0 (C-Csp2) 285.73 (C-O)	6513 4691	1.33 1.40	10,441 7432	58.4 41.6

As shown in Table 2, the FWHM of the two peaks of sample h does not quite match. This is because, considering that the C-C on the surface of carbon fibers mostly exist in the form of SP2 hybridization, and due to the π-π excitation effect, most of the peaks of carbon fibers have a long tail. Therefore, a nonlinear fitting method was chosen during the fitting process. This led to the FWHM of the two peaks of sample h not matching well.

3.4. The Tensile Strength of Fibers

Six specimens of each sample were tested using a universal material testing machine (INSTRON 5966) at a loading rate of 5 mm/min, and the average tensile strengths of the exposed and sample h were obtained, as shown in Figure 11 The tensile strength of the blank sample was 5.74 GPa, and the standard deviation was 0.257 GPa, which is in line with the tensile strength of T800 grade carbon fiber multifilament given by the manufacturer. After LPE, the tensile strength of the LPE Sample was 4.28 GPa, and the standard deviation was 0.292 GPa. After LPE treatment, the tensile strength of the fibers was retained by 75%. This might be due to the bombardment effect and surface etching effect of the liquid-phase plasma, which locally damaged the surface of the carbon fiber, resulting in the obstruction of load transfer and ultimately leading to the decrease in the tensile strength of the carbon fiber.

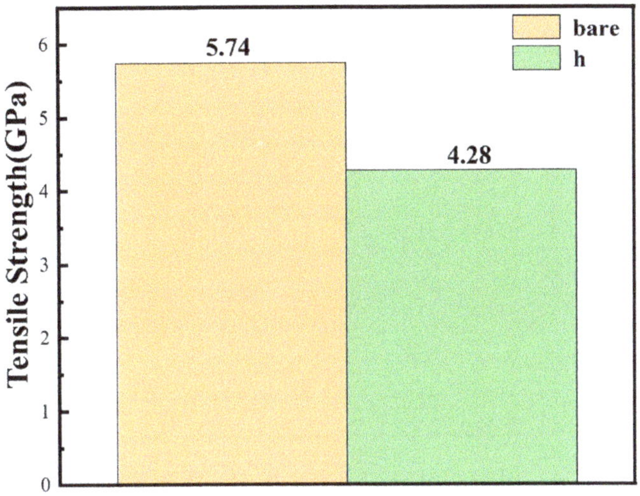

Figure 11. The tensile strength of the bare sample and sample h.

3.5. Mechanism of LPE in Removing Sizing Agents from the Surface of Carbon Fibers

The main components of the sizing agent on the surface of carbon fibers include water-soluble and non-water-soluble epoxy resins, dimethyl resin, polyimide resin, polyvinyl alcohol resin, vinyl acetate resin, acrylic resin, etc. The water-soluble sizing agent components dissolve when passing through the aqueous solution, while the non-water-soluble sizing agent has a melting point between 100 °C and 180 °C and a flash point between 225 °C and 235 °C.

Figure 12 reveals the mechanism of removing sizing agents by liquid-phase plasma. When the carbon fiber passes beneath the cathode, the resin on it may first melt under the influence of the high-temperature heat-affected zone of the electrode itself and the plasma. Furthermore, due to its insolubility in water and its density being lower than that of water, the molten sizing agent may separate from the carbon fiber due to the force generated during the high-speed movement of the fiber in the solution and the bursting of bubbles, rise to the plasma arc zone, and be ignited and vaporized. The process is shown in Figure 12a. Meanwhile, as the carbon fiber is relatively close to the cathode, the potential of the carbon fiber remains low enough throughout the electric field from the cathode to the anode to act as a false cathode. So, during the electrification process, high-voltage bubbles are constantly generated and burst on the cathode and the carbon fiber, and the surface of the carbon fiber is subjected to mechanical vibration disturbed by the bubbles in the solution. The work of Zhang et al. [34] also demonstrated the fact that carbon fibers are polarized as conductors in an electric field. This is conducive to the separation of sizing agents with fluid properties. The process is shown in Figure 12b. According to the bubble discharge theory, as high-voltage bubbles grow and burst, the internal pressure of the bubbles decreases, inducing gas discharge within the bubbles. This leads to the destruction of liquid insulation, the breakdown of the bubbles, and the generation of plasma. The generated plasma exerts a more intense mechanical bombardment and stripping effect on the surface of carbon fibers. This promotes the separation of sizing agents. The research of Gupta et al. [35] is the origin of this theory that triggers liquid-phase plasma. The process is shown in Figure 12c. Some components in the sizing agent are water-soluble. After dissolving in water, the ionized cations and anions will be affected by the strong electric field in the plasma arc region and separate towards the anode and cathode. This promotes the decomposition of the sizing agent components. The process is shown in Figure 12d.

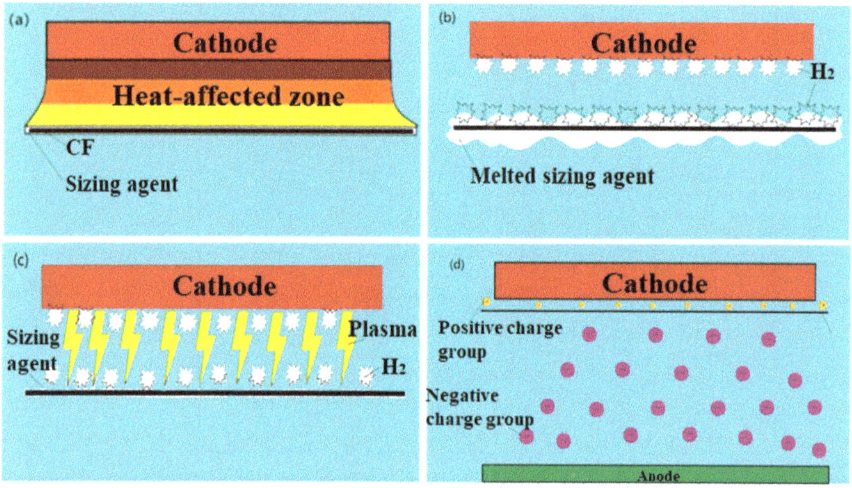

Figure 12. Schematic diagram of the mechanism of LPE in removing the sizing agent, (**a**) The thermal field effect during LPE, (**b**) The mechanical force acting on the generation and rupture of bubbles in LPE, (**c**) The direct bombardment effect of plasma in LPE, (**d**) The water-soluble components in the sizing agent during LPE are affected by the fluid field and the electric field.

4. Conclusions

In this study, the sizing agent on the surface of T800 carbon fiber was successfully removed by liquid-phase plasma electrolysis (LPE). By changing parameters such as different arc voltages (185–215 V) and electrode spacings (10–20 mm). We obtained a series of samples. The possible range of the sample with the optimal parameters was estimated through the SEM images. The TGA and XPS analyses and the comparison of samples with the high-temperature removal of sizing agent confirmed that the combination of 200 V and 10 mm spacing could almost completely remove the sizing agent. Under this condition, plasma bombardment dominated the removal mechanism, eliminating sizing residues while exposing the underlying fiber grooves. TGA further demonstrated that samples treated at 10 mm spacing exhibited negligible weight loss, indicating complete sizing removal. The results of XPS further confirmed the high efficiency of LPE in the removal of sizing agents (C-O content from 41.6% to 22.4%), and the retention of C-O also proved that LPE could maintain the surface activity of carbon fibers, confirming the effectiveness of LPE in decomposing the sizing agent. Meanwhile, based on the above test results, an attempt was made to explain the mechanism of LPE in removing sizing agents from the surface of carbon fibers. This method offers a rapid and solvent-free alternative that can replace traditional de-gumming techniques, as well as the pretreatment steps before large-scale fiber surface modification treatment or the cleaning of fiber surfaces during the production process. It has potential advantages in the manufacturing of carbon fiber composite materials. Future work should explore the possibility of simultaneous sizing agent removal and coating deposition on the fiber surface of LPE and its influence on adhesion at the fiber–matrix interface in composite materials.

Author Contributions: Conceptualization, C.H.; methodology, C.H.; validation, X.H., D.L. and Q.Z.; investigation, W.C.; resources, M.L.; data curation, X.W.; writing—original draft preparation, C.H.; writing—review and editing, Q.Z.; visualization, C.H.; supervision, W.C.; project administration, W.C.; funding acquisition, M.L. All authors have read and agreed to the published version of the manuscript.

Funding: This work was supported by the National Natural Science Foundation of China [grant numbers: U2341262].

Data Availability Statement: The data that support the findings of this study are available from the corresponding author, W.C., upon reasonable request.

Conflicts of Interest: The authors declare no conflict of interest.

References

1. Savino, R.; Criscuolo, L.; Di Martino, G.D.; Mungiguerra, S. Aero-thermo-chemical characterization of ultra-high-temperature ceramics for aerospace applications. *J. Eur. Ceram. Soc.* **2018**, *38*, 2937–2953. [CrossRef]
2. Tiwary, A.; Kumar, R.; Chohan, J.S. A review on characteristics of composite and advanced materials used for aerospace applications. *Mater. Today Proc.* **2022**, *51*, 865–870. [CrossRef]
3. Wang, X.; Gao, X.; Zhang, Z.; Cheng, L.; Ma, H.; Yang, W. Advances in modifications and high-temperature applications of silicon carbide ceramic matrix composites in aerospace: A focused review. *J. Eur. Ceram. Soc.* **2021**, *41*, 4671–4688. [CrossRef]
4. Voronina, S.Y.; Shalygina, T.A.; Voronchikhin, V.D.; Vlasov, A.Y.; Ovchinnikov, A.N.; Grotskaya, N.N. Data for determining the surface properties of carbon fiber in contact interaction with polymeric binders. *Data Brief* **2021**, *35*, 106847. [CrossRef]
5. Albano, M.; Alifanov, O.M.; Budnik, S.A.; Morzhukhina, A.V.; Nenarokomov, A.V.; Titov, D.M.; Gabrielli, A.; Ianelli, S.; Marchetti, M. Carbon/carbon high thickness shell for advanced space vehicles. *Int. J. Heat Mass Transf.* **2019**, *128*, 613–622. [CrossRef]
6. Du, C.; Zhang, X.; Zhou, R.; Sang, Z.; Jiang, Y. Multi-scale modeling and performance study of carbon fiber heterogeneous electrode batteries based on electrochemical-mechanical coupling. *J. Power Sources* **2025**, *648*, 237404. [CrossRef]
7. Salimkhani, H.; Palmeh, P.; Khiabani, A.B.; Hashemi, E.; Matinpour, S.; Salimkhani, H.; Asl, M.S. Electrophoretic deposition of spherical carbonyl iron particles on carbon fibers as a microwave absorbent composite. *Surf. Interfaces* **2016**, *5*, 1–7. [CrossRef]
8. Movassagh-Alanagh, F.; Bordbar-Khiabani, A.; Ahangari-Asl, A. Fabrication of a ternary PANI@Fe3O4@CFs nanocomposite as a high performance electrode for solid-state supercapacitors. *Int. J. Hydrogen Energy* **2019**, *44*, 26794–26806. [CrossRef]
9. Shin, S.; Jang, J.; Yoon, S.H.; Mochida, I. A study on the effect of heat treatment on functional groups of pitch based activated carbon fiber using FTIR. *Carbon* **1997**, *35*, 1739–1743. [CrossRef]
10. Fanning, P.E.; Vannice, M.A. A DRIFTS study of the formation of surface groups on carbon by oxidation. *Carbon* **1993**, *31*, 721–730. [CrossRef]
11. Lee, S.; Ham, S.; Youn, S.J.; Chung, Y.S.; Lee, S. Effect of Textile PAN-Based Carbon Fibers with Rough Surface on Interfacial Adhesion in PA6 Composites. *Ind. Eng. Chem. Res.* **2021**, *60*, 9088–9097. [CrossRef]
12. Wang, H.; Jin, K.; Tao, J. Improving the interfacial shear strength of carbon fibre and epoxy via mechanical interlocking effect. *Compos. Sci. Technol.* **2020**, *200*, 108423. [CrossRef]
13. Zhu, P.; Shi, J.; Bao, L. Effect of polyetherimide nanoparticle coating on the interfacial shear strength between carbon fiber and thermoplastic resins. *Appl. Surf. Sci.* **2020**, *509*, 145395. [CrossRef]
14. Fukunaga, A.; Ueda, S. Anodic surface oxidation for pitch-based carbon fibers and the interfacial bond strengths in epoxy matrices. *Compos. Sci. Technol.* **2000**, *60*, 249–254. [CrossRef]
15. Tran, M.Q.; Ho, K.K.C.; Kalinka, G.; Shaffer, M.S.P.; Bismarck, A. Carbon fibre reinforced poly(vinylidene fluoride): Impact of matrix modification on fibre/polymer adhesion. *Compos. Sci. Technol.* **2008**, *68*, 1766–1776. [CrossRef]
16. Zhang, Y.; Xu, F.; Zhang, C.; Wang, J.; Jia, Z.; Hui, D.; Qiu, Y. Tensile and interfacial properties of polyacrylonitrile-based carbon fiber after different cryogenic treated condition. *Compos. Part B Eng.* **2016**, *99*, 358–365. [CrossRef]
17. Sharma, M.; Gao, S.; Mäder, E.; Sharma, H.; Wei, L.Y.; Bijwe, J. Carbon fiber surfaces and composite interphases. *Compos. Sci. Technol.* **2014**, *102*, 35–50. [CrossRef]
18. Lee, S.; Hoang, Q.N.; Lee, H.; Chung, Y.S.; Park, H.; Lee, S. Effect of plasma modification on the anisotropic surface structure of PAN-based graphitic carbon fiber. *Carbon* **2025**, *242*, 120429. [CrossRef]

19. Lee, S.-C.; White, S.; Grzesik, J.A. Effective radiative properties of fibrous composites containing spherical particles. *J. Thermophys. Heat Transf.* **1994**, *8*, 400–405. [CrossRef]
20. Fang, H.; Sheng, Z.; Wang, W.; Wei, C.; Li, S.; Geng, X.; Li, X.; Zhu, N.; Wen, G.; Dong, S.; et al. Formation and mechanism of carbon coating on carbon fibers through glucose-to-carbon conversion and its effect on the mechanical properties of Cf/ZrB2-SiC composites. *J. Eur. Ceram. Soc.* **2025**, *45*, 117569. [CrossRef]
21. Patra, N.; Ramesh, P.; Ravi, G.; Ţălu, Ş. Electroless deposition of nanoscale Ni-Co-Fe-P coatings on carbon fibers for multifunctional applications. *Diam. Relat. Mater.* **2025**, *158*, 112602. [CrossRef]
22. Shao, Y.; Xu, F.; Liu, W.; Zhou, M.; Li, W.; Hui, D.; Qiu, Y. Influence of cryogenic treatment on mechanical and interfacial properties of carbon nanotube fiber/bisphenol-F epoxy composite. *Compos. Part B Eng.* **2017**, *125*, 195–202. [CrossRef]
23. Zafeiropoulos, N.E.; Baillie, C.A.; Hodgkinson, J.M. Engineering and characterisation of the interface in flax fibre/polypropylene composite materials. Part II. The effect of surface treatments on the interface. *Compos. Part A Appl. Sci. Manuf.* **2002**, *33*, 1185–1190. [CrossRef]
24. Kwon, D.W.; Nam, K.B.; Hong, S.C. Influence of tungsten on the activity of a Mn/Ce/W/Ti catalyst for the selective catalytic reduction of NO with NH3 at low temperatures. *Appl. Catal. A Gen.* **2015**, *497*, 160–166. [CrossRef]
25. Li, H. Efficient reduction of L-infinity geometry problems. In Proceedings of the 2009 IEEE Conference on Computer Vision and Pattern Recognition, CVPR 2009, Miami, FL, USA, 20–25 June 2009; pp. 2695–2702.
26. Donnet, J.B.; Brendle, M.; Dhami, T.L.; Bahl, O.P. Plasma treatment effect on the surface energy of carbon and carbon fibers. *Carbon* **1986**, *24*, 757–770. [CrossRef]
27. Shi, D.; Lian, J.; He, P.; Wang, L.M.; Xiao, F.; Yang, L.; Schulz, M.J.; Mast, D.B. Plasma coating of carbon nanofibers for enhanced dispersion and interfacial bonding in polymer composites. *Appl. Phys. Lett.* **2003**, *83*, 5301–5303. [CrossRef]
28. Lu, P.; Zhang, Z.; Yang, M.; Wu, J.; Chen, L.; Xue, W. Efficient synthesis and lithium storage performance of SiO2/TiO2 composite film anode by plasma electrolytic oxidation. *Mater. Lett.* **2024**, *371*, 136902. [CrossRef]
29. Bokhari, S.W.; Wei, S.; Gao, W. Synthesis of bimetallic MoS2/VS2 nano-urchins-reduced graphene oxide hybrid nanocomposite for high performance supercapacitor application. *Electrochim. Acta* **2021**, *398*, 139300. [CrossRef]
30. Lee, E.-s.; Lee, C.-h.; Chun, Y.-S.; Han, C.-j.; Lim, D.-S. Effect of hydrogen plasma-mediated surface modification of carbon fibers on the mechanical properties of carbon-fiber-reinforced polyetherimide composites. *Compos. Part B Eng.* **2017**, *116*, 451–458. [CrossRef]
31. Xiao, J.; Zhang, X.; Zhao, Z.; Liu, J.; Chen, Q.; Wang, X. Rapid and Continuous Atmospheric Plasma Surface Modification of PAN-Based Carbon Fibers. *ACS Omega* **2022**, *7*, 10963–10969. [CrossRef] [PubMed]
32. Moosburger-Will, J.; Bauer, M.; Schubert, F.; Kunzmann, C.; Lachner, E.; Zeininger, H.; Maleika, M.; Hönisch, B.; Küpfer, J.; Zschoerper, N.; et al. Methyltrimethoxysilane plasma polymerization coating of carbon fiber surfaces. *Surf. Coat. Technol.* **2017**, *311*, 223–230. [CrossRef]
33. *GB/T 29761-2022*; Carbon Fibre—Determination of Sizing Content. 7.2 Method B: High-Temperature Decomposition Method. National Standardization Administration Committee, State Administration for Market Regulation: Beijing, China, 2022.
34. Zhang, Y.; Chen, C.; Chen, W.; Cheng, H.; Wang, L. A novel aqueous plasma electrolysis for carbon fiber. *Chem. Eng. J.* **2016**, *304*, 426–430. [CrossRef]
35. Gupta, P.; Tenhundfeld, G.; Daigle, E.O.; Ryabkov, D. Electrolytic plasma technology: Science and engineering—An overview. *Surf. Coat. Technol.* **2007**, *201*, 8746–8760. [CrossRef]

colloids and interfaces

Article

Enhanced Recovery of an Arsenopyrite-Type Gold Ore: Flotation Surface Chemistry and Kinetics of Blended Collector W8 with ADD

Qingqing Xing [1,2], Fei Li [1,2,*], Pingtian Ming [1,2,3] and Zhen Wang [4,*]

1 The Sixth Geological Exploration Institute of Qinghai Province, Xining 810029, China; ramble77077@sina.com (Q.X.); ptming4500@sina.com (P.M.)
2 Qinghai Engineering Research Center for Gold Mineral Resource Development Dressing and Metallurgy Pilot Plant, Dulan Jin Hui Mining Limited Corporation, Haixi Zhou 816100, China
3 School of Civil and Resource Engineering, University of Science and Technology Beijing, Beijing 100089, China
4 School of Environment and Resource, Southwest University of Science and Technology, Mianyang 621010, China
* Correspondence: a376169069@163.com (F.L.); wangzhen@swust.edu.cn (Z.W.)

Abstract

This study investigated the flotation performance of W8, a blended xanthate collector containing ethyl, butyl, propyl, and amyl xanthates, combined with ammonium dibutyl dithiophosphate (ADD) for treating low-grade arsenopyrite-type gold ore from Golmud, Qinghai. Real ore flotation tests demonstrated the superior efficacy of the W8 + ADD system, achieving 84.06% gold recovery with 0.34 g/t tailings, outperforming conventional sodium amyl xanthate (SAX) + ADD and sodium propyl xanthate (SPX) + ADD systems. Systematic studies on pure arsenopyrite revealed a significant synergistic effect in the mixed SPX-SAX system (1:4 ratio), representative of W8 composition. At pH 9, the mixed collector achieved 73.5% recovery, substantially higher than individual SPX (37.5%) or SAX (45.8%). This enhanced performance was attributed to improved surface hydrophobicity (contact angle 47.68° vs. 36.92° for SAX), greater adsorption density (4.97×10^{-7} mol/g under depressant conditions), and extensive formation of molecular aggregates observed via AFM, which increased surface roughness to 28.95 nm. Flotation kinetics further confirmed the advantage of W8 + ADD, which reached 72.1% cumulative recovery in 420 s, exceeding both mixed SPX/SAX (69.5%) and single SAX (65.5%) systems. The synergistic interaction among different xanthate components in W8 enables efficient recovery of gold from this refractory ore.

Keywords: collector; arsenopyrite; gold ore; synergistic effect; flotation kinetics

1. Introduction

Gold is a strategic metal with unique properties and extensive applications [1,2]. With the rapid growth in domestic gold demand and the current resource landscape in China, the exploitation of low-grade and associated/disseminated gold deposits has become an inevitable necessity. Among the proven gold reserves in China, associated gold accounts for 30% of the total reserves [3,4]. Taking the Wulonggou Gold Mine in Dulan County, Qinghai Province as an example, it is a representative gold deposit of this type in the Qaidam Basin. The mining area consists of three independent ore sections, covering a total area of 2.12 square kilometers, with a total resource reserve of 120.81 tons and an estimated potential economic value exceeding 10 billion American dollars. The majority of China's associated gold deposits are arsenopyrite-type, primarily distributed in the northwestern region. These are low-grade polysulfide gold ores characterized by fine dissemination of gold-bearing minerals, often associated with clay minerals such as sericite, chlorite, and kaolinite; sulfides with poor floatability like arsenopyrite serve as the primary gold-bearing minerals, making these ores typically complex and refractory [5,6].

The efficient flotation of refractory minerals imposes higher demands on the performance of flotation reagents, particularly collectors. The development of high-performance collectors that combine strong collecting power with high selectivity remains a key research focus and challenge in the flotation field [7]. Two main strategies are currently pursued: first, the molecular design of novel collectors [8]. For instance, by introducing heteroatoms or heterocyclic groups into traditional xanthate molecules, the strength and nature of interactions between functional groups and surface-active atoms on minerals can be modulated to balance selectivity and collecting capacity, thereby enabling the design of new, highly effective collectors [9]. Second, the blended use of existing collectors [10]. In cases of complex mineral intergrowth, a single collector often fails to achieve satisfactory separation. However, the combination of reagents according to certain principles can produce unexpected synergistic effects—commonly referred to as synergism. Compared to single reagents, blended reagent systems can improve flotation efficiency, increase concentrate grade, reduce reagent consumption, or optimize the flotation environment [11,12].

For instance, the use of a mixed collector system comprising octyl hydroxamic acid (OHA) and sodium dodecyl sulfonate (SDS) enhanced columbite recovery while reducing operational costs [13]. To address the issue of high reagent costs at a copper smelting slag processing plant, Wang Zitao et al. conducted experimental studies with various collectors. Their results demonstrated that a combined collector system of Z-200 and butyl xanthate provided superior performance, optimizing metallurgical performance while simultaneously reducing reagent consumption [14]. In a case study conducted by Bradshaw, synergies between thiol collectors during pyrite flotation at pH 4 were reported. Potassium butyl xanthate and a dithiocarbamate collector were evaluated both as individual collectors and as components of a blended collector system. Batch flotation tests assessed the impact of the mixed collector on overall flotation performance by measuring sulfur recovery and grade, water recovery, and flotation kinetics. The results indicated that the use of the collector mixture enhanced the bubble-particle collection efficiency [15].

W8 is a blended xanthate collector comprising ethyl, butyl, propyl, and amyl xanthates. In this study, it was applied to the flotation of a gold ore from Golmud, Qinghai, where arsenopyrite serves as the primary gold-bearing mineral [16]. At present, under the collector system composed of amyl xanthate and butylamine black drug used on site, the recovery rate of gold concentrate is not good. Preliminary tests revealed that W8 outperformed pure amyl xanthate, even when both were used in combination with ammonium dibutyl dithiophosphate. Building on previous findings by other researchers, the enhanced mechanism of W8 was systematically investigated through a series of methods including micro-flotation tests on pure minerals, contact angle measurements, adsorption density analysis, surface morphology characterization, and flotation kinetics studies.

2. Materials and Methods

2.1. Materials and Reagents

The raw ore samples were collected from the Wulonggou mining area in Golmud City, Qinghai Province. The element composition is shown in Table 1. The gold ore assayed 2.16 g/t, a submarginal grade that remains economically viable for extraction under current technological and economic conditions [17]. Chemical analysis revealed a composition of 62.25% SiO_2 and 13.37% Al_2O_3, indicating a high content of quartz and silicate/aluminosilicate minerals, which constitute the primary acidic gangue. Gold is mainly encapsulated in carrier minerals in a micro-submicroscopic state, accounting for more than 70% (with the remaining less than 30% distributed in other minerals): approximately 39.11% of the gold is encapsulated in arsenopyrite, 11.01% in pyrite, and 1.69% in pyrrhotite. Furthermore, the elevated concentrations of arsenic and carbon categorize this ore as refractory, presenting significant challenges for conventional cyanidation. The focus of this study is not on the gangue minerals, but on the efficient flotation of the gold-bearing mineral, arsenopyrite.

Table 1. Chemical multi-element analysis results of the ore.

Element	Au *	Ag *	S	As	SiO_2	Al_2O_3	CaO
Content/%	2.16	1.54	2.01	0.26	62.25	13.37	3.52
Element	**MgO**	**Cu**	**Zn**	**Ni**	**Co**	**TFe**	**C**
Content/%	1.90	0.008	0.056	0.007	0.002	4.10	0.38

* Unit of Au/Ag content, g/t.

Pure minerals of arsenopyrite were obtained from Beijing Shuiyuan Shanchang mineral specimen Co., LTD. (Beijing, China). The samples were dry-ground using a jar mill with porcelain balls and subsequently classified by sieving through screens with apertures of 0.045 and 0.026 mm, respectively. The −0.045 + 0.026 mm fraction was reserved for micro-flotation tests, adsorption studies and contact angle measurements. This is a relatively common method for mineral sample processing [18]. Laser particle size analysis was conducted on the samples obtained through sieving, and the resulting curves are shown in Figure 1. It is obvious that the particle size of the samples is indeed within the expected range. The phase purity of arsenopyrite was verified by X-ray diffraction (XRD; X'Pert PRO, PANalytical B.V., Almelo, The Netherlands) to obtain the spectrum (Figure 2).

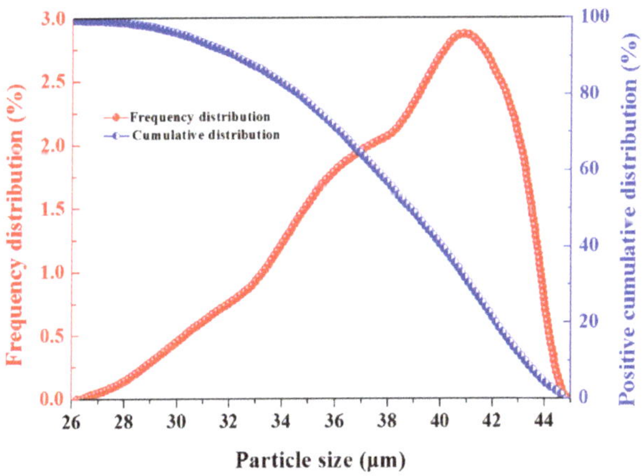

Figure 1. The particle size distribution of the sieved arsenopyrite sample.

Analytically pure hydrochloric acid and sodium carbonate Na_2CO_3 were used as the pH regulators. Analytically pure terpilenol ($C_{10}H_{18}O$ (4-methyl-2-pentanol)) was used as the frother. Copper sulfate ($CuSO_4$), sodium amyl xanthate (SAX, $C_5H_{11}OCSSNa$), sodium propyl xanthate (SPX, $C_3H_7OCSSNa$), and Ammonium dibutyl dithiophosphate (ADD,$(C_4H_9O)_2PSSNH_4$), were also analytical reagent grade and used as the collector. The above analytically pure reagents were bought from Sinopharm Group (Beijing, China).

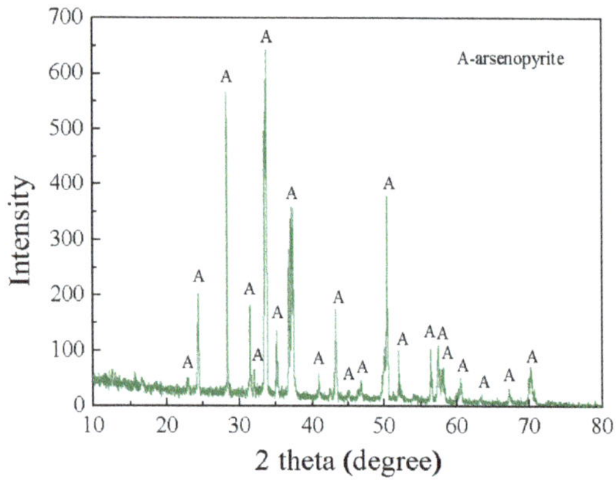

Figure 2. XRD patterns of arsenopyrite used in the tests.

2.2. Flotation Tests

2.2.1. Flotation of the Real Ore

The open circuit flow sheet for collector comparison with one roughing and one scavenging is shown in Figure 2. A total of 500 g of raw ore was ground to −0.074 mm size fractions accounting for 78% in a laboratory conical ball mill and placed into the 1.5 L flotation cell (XFD-type flotation apparatus), and the flotation cell was filled with water to prepare a pulp with the solid concentration of approximately 40 w.%. The slurry pH was regulated using Na_2CO_3 solution; $CuSO_4$ was used as the activator of Au-bearing sulfide minerals; SWUST-D01 (polysaccharide) was employed as the depressant; SAX + ADD or W8 + ADD was used as the collector; and 2# oil (industrial products of terpineol) was used as the frother. The condition time for each step and the reagent dosage are shown in Figure 3. After finishing the roughing operation, the scavenging was conducted with lower reagent dosage. The floated fraction of roughing was the concentrate, and that of scavenging was the middlings; the product remaining in the flotation cell was the tailings. The final concentrate, middlings, and tailings were filtered, dried, weighed, and analyzed for Au.

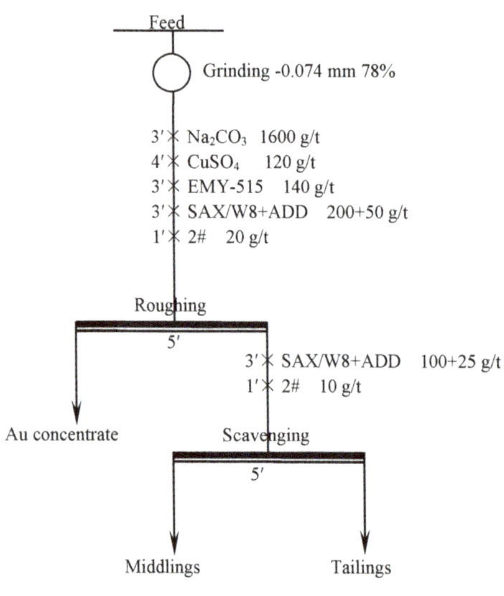

Figure 3. Flow chart of open circuit flotation test for collector comparison.

2.2.2. Flotation of Pure Mineral

Micro-flotation tests on pure minerals were performed using a 40 mL flotation cell. In each test, 2.0 g of the mineral sample was charged into the cell containing 35 mL of distilled water and agitated at 1920 r/min for 2 min. The pH was then adjusted to the desired value using HCl or Na_2CO_3 as regulators. Following pH conditioning, the activator was introduced, followed by the collector, with each reagent addition succeeded by 3 min of conditioning. Subsequently, terpinenol was added and the pulp was agitated for another 2 min prior to the 5-min flotation process. Both the floated and non-floated products were collected, separately filtered, dried, and weighed. The flotation recovery was calculated based on the dry weights of the concentrates and tailings. All tests were conducted in triplicate under identical conditions, and the average recovery was reported, with the deviation among replicates maintained within 3%.

For the flotation kinetics study, a single dose of reagents was added initially, followed by the collection of froth products in six timed intervals: the first two at 30 s each, the next two at 60 s each, and the final two at 120 s each.

2.3. Contact Angle

To determine the wettability of the arsenopyrite surface, a contact angle measurement apparatus was employed in this study. First, mineral tablets of the arsenopyrite powder were prepared, and then treated with various collectors. Subsequently, these treated bulk samples were air-dried naturally at room temperature to ensure the consistency of surface conditions. For the contact angle measurement, the sessile drop method was adopted to determine the contact angle of the processed thin tablets.

2.4. Adsorption Measurements

The adsorption density of xanthate on arsenopyrite was determined by measuring the absorbance at the characteristic wavelength of 301 nm using a UV-3100 spectrophotometer. In a typical procedure, a 2.0 g sample of the −0.045 + 0.026 mm fraction was pulped with 40 mL of distilled water, followed by the addition of specified reagents. The slurry was agitated for 15 min and subsequently centrifuged for 10 min at 4500 rpm using a high-speed refrigerated centrifuge (S-1-150s, HONGHUAYIQI Co., Ltd., Cangzhou, China). The absorbance of the resulting supernatant was then measured. The residual xanthate concentration was determined by referring to a pre-established calibration curve (Figure 4). Finally, the amount of xanthate adsorbed on the mineral surface was calculated based on the difference between the initial and equilibrium concentrations [19].

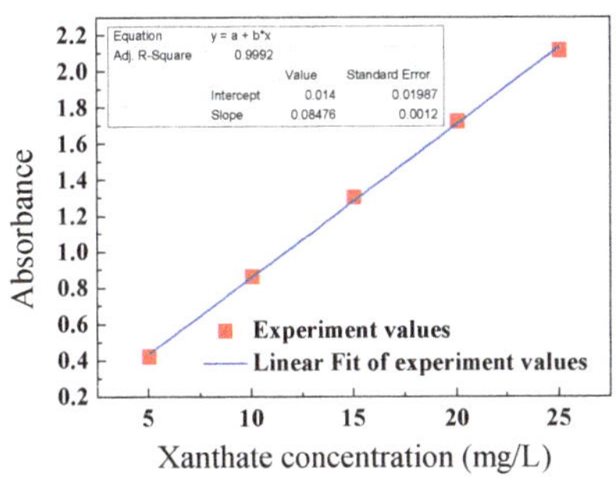

Figure 4. The standard curve of xanthate.

2.5. Atomic Force Microscopy Observation

The surface morphology of arsenopyrite before and after collector adsorption was characterized in situ using atomic force microscopy (AFM; Dimension Icon, Bruker Nano

Inc., USA). Sample preparation involved sequential wet polishing with 600-, 800-, 1000-, 1200-, 1500-, 2000-, and 2500-mesh abrasive papers, followed by final polishing with diamond suspensions of 3 μm, 1 μm, and 0.2 μm particle sizes [20]. After being immersed in the respective collector solutions for 6 h, the polished sections were thoroughly rinsed with deionized water, dried under a gentle nitrogen stream, and subsequently examined by AFM. Imaging was conducted under ambient conditions using ScanAsyst mode for topographical analysis and force modulation mode for mechanical property mapping, with antimony (n)-doped silicon cantilevers (TESPA-V2, Bruker, Billerica, MA, USA).

3. Results and Discussion

3.1. Real Ore Flotation

The separation indicators of a certain gold ore in Golmud under three reagent systems were comparatively studied, and the results are shown in Table 2. In the production plant of a Golmud gold mine, the combined collector system of SAX and ADD was employed. When applied to the experimental flowsheet, this system achieved a high gold concentrate grade of 14.10 g/t, along with a recovery of 81.94%, while reducing the gold content in the tailings to 0.37 g/t. These results are considered satisfactory. For comparison, when SPX was used to replace SAX at the same dosage, forming an SPX + ADD collector system, the resulting gold recovery and concentrate grade were 81.27% and 13.4 g/t, respectively. Both of these key indicators were lower than those obtained with the SAX + ADD system. Obviously, this is related to the longer chain length of the hydrocarbons in SAX [21].

Table 2. Test results of collector comparison.

Type of Collector and the Dosage Selected for Roughing and Scavenging (g/t)	Product	Yield (%)	Au Grade (g/t)	Au Recovery (%)
SPX + ADD (200 + 50)/(100 + 25)	Au Concentrate	13.05	13.40	81.27
	Middlings	2.76	2.76	3.54
	Tailings	83.79	0.39	15.19
	Raw Ore	99.60	2.16	100.00
SAX + ADD (200 + 50)/(100 + 25)	Au Concentrate	13.04	14.10	81.94
	Middlings	3.72	2.64	4.38
	Tailings	82.94	0.37	13.68
	Raw Ore	99.70	2.25	100.00
W8 + ADD (200 + 50)/(100 + 25)	Au Concentrate	13.88	13.30	84.06
	Middlings	2.48	2.65	3.00
	Tailings	83.64	0.34	12.94
	Raw Ore	100.00	2.20	100.00

When the novel collector W8 was substituted for SAX, the new W8 + ADD system yielded the best overall performance: a gold recovery of 84.06% and a concentrate grade of 13.30 g/t, with the gold content in tailings further reduced to 0.34 g/t. Given that recovery is generally of greater practical significance than grade during the roughing and scavenging stages, the W8 + ADD collector system clearly delivered the most favorable outcomes [22]. It demonstrates the superior flotation performance of W8 + ADD for this specific gold ore compared to the collector system currently used in the plant.

3.2. Micro Flotation of Pure Mineral

The novel collector W8 has a complex composition, being a mixture of several xanthates and dithiophosphates in specific proportions. To simplify the experimental system and more directly reveal the physicochemical mechanism underlying the synergistic effects of reagent blending, this study investigates the flotation behavior and interfacial interactions of arsenopyrite using a mixed propyl/amyl xanthate collector system. The results are shown in Figure 5.

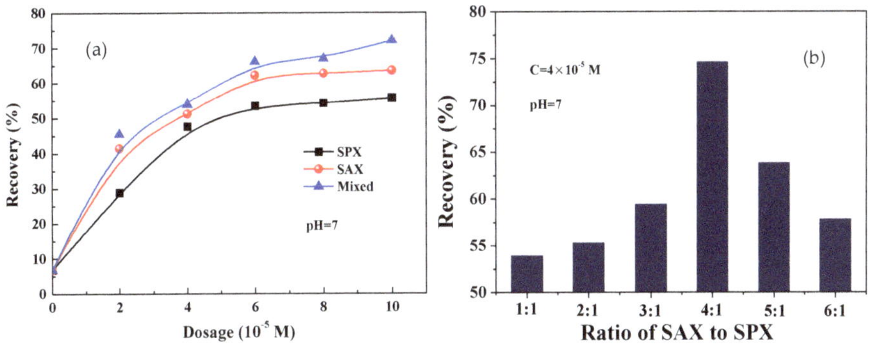

Figure 5. The relationship between the flotation recovery of arsenopyrite (**a**) in different collector solution and the concentration of reagent (for the mixed system SAX: SPX = 1:1; pH 7), and (**b**) in a mixed collector system and the mixing ratio.

When used individually, SAX generally demonstrated better collecting power than SPX. Both collectors reached their maximum flotation efficiency at a dosage of 6×10^{-5} mol/L. Further increasing the dosage did not significantly improve the recovery of arsenopyrite, which is likely attributed to the approach of monolayer adsorption saturation on the mineral surface at higher concentrations [23]. Beyond this point, the incremental increase in adsorption density decreases sharply.

When SPX and SAX were used in combination (Mixed, mass ratio 1:1), a noticeable improvement in recovery was observed. That is, replacing part of the long-chain SAX with short-chain SPX resulted in better flotation performance than using SAX alone. This finding contrasts with the conventional understanding that longer-chain collectors typically exhibit superior performance when used individually and instead indicates a synergistic interaction between short-chain SPX and long-chain SAX in the mixed system, enhancing the flotation recovery of arsenopyrite [24].

Figure 6 illustrates the relationship between pulp pH and the flotation recovery of arsenopyrite in different xanthate systems. Overall, the flotation recovery of arsenopyrite decreased with increasing pulp pH across all reagent systems. However, at any given pH value, the mixed collector system consistently yielded significantly higher recovery compared to individual collectors. For instance, under weakly alkaline conditions (pH $\approx$ 9), the recoveries achieved with SPX, SAX, and the mixed system were 37.5%, 45.8%, and 73.5%, respectively, indicating a pronounced enhancement in flotation performance with the mixed collectors.

Furthermore, it can be observed that when the pH exceeded the neutral range, the flotation recovery of arsenopyrite declined rapidly in single-collector systems. In contrast, with the mixed collector system, the recovery remained relatively stable from the neutral pH range up to around pH 9. This suggests that the mixed collectors not only enhance the adsorption capacity on the arsenopyrite surface but also exhibit stronger adaptability to pH variations compared to individual collector systems.

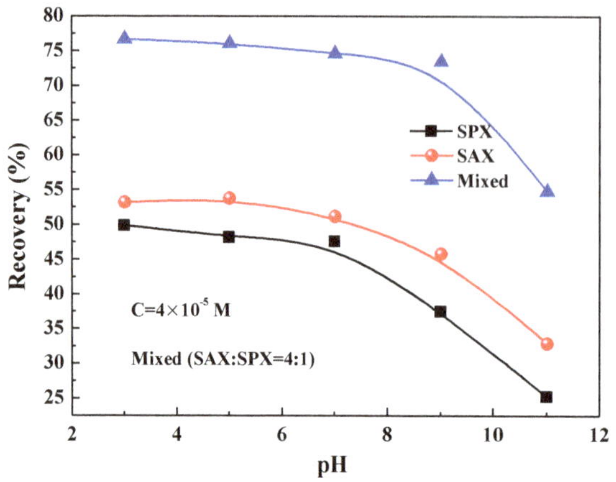

Figure 6. The relationship between the flotation recovery of arsenopyrite in different collector solution systems and the pH of the pulp (for the mixed system SAX-SPX = 4:1; C = 4×10^{-5} mol/L).

3.3. Contact Angle Measurement

Measuring the contact angle is a widely used method to assess changes in surface wettability [25]. A smaller contact angle indicates better wettability (i.e., poorer hydrophobicity), while a larger angle indicates poorer wettability (i.e., better hydrophobicity). The flotation process relies heavily on the wettability of specific mineral surfaces in the pulp. For instance, the adsorption of collectors onto valuable mineral surfaces inevitably reduces their wettability, whereas the interaction of depressants with gangue minerals increases it [26]. Thus, contact angle measurements performed on arsenopyrite—a characteristic gold-bearing mineral in this ore—before and after reagent treatment, help clarify the performance of the reagents.

Figure 7 presents the contact angle results for arsenopyrite before and after interaction with the reagent. In this study, contact angles were measured on pressed powder pellets of arsenopyrite, which explains the difference from values reported in the literature for natural arsenopyrite surfaces. Nevertheless, the comparison of contact angles before and after reagent treatment effectively reveals the reagents' impact and their effect on mineral surface hydrophilicity/hydrophobicity [27].

Figure 7a shows the arsenopyrite surface without any reagent treatment. Figure 7b–d show the surfaces after treatment with the collector SPX, SAX, and the mixed-xanthate system, respectively.

As shown in the figures, the contact angle on the untreated arsenopyrite surface is 17.98°, a relatively low value indicating its naturally hydrophilic. After treatment with SPX, the contact angle increases significantly to 33.54°, demonstrating that this collector substantially enhances the hydrophobicity of the arsenopyrite surface. Treatment with SAX results in a further increase in the contact angle to 36.92°, indicating its stronger ability to impart hydrophobicity compared to SPX. This observation aligns with the literature reports that longer-chain xanthate collectors generally exhibit greater collecting power and hydrophobization capability [28].

Notably, after treatment with the mixed-xanthate system, the contact angle reaches 47.68°, the highest value observed. This suggests that the mixed collector system has a stronger hydrophobizing effect on the arsenopyrite surface than SAX alone.

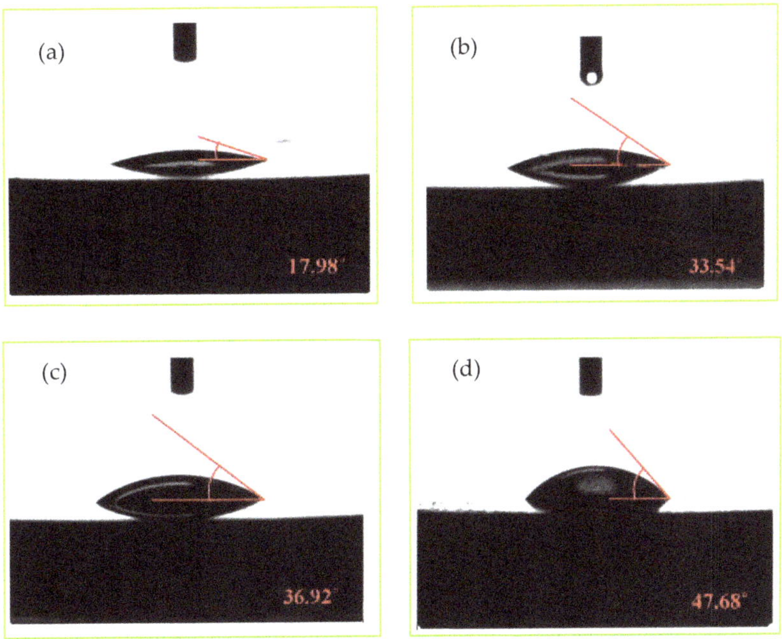

Figure 7. The surface contact angle of the arsenopyrite pressed powder after its action in (**a**) water, (**b**) SPX, (**c**) SAX, and (**d**) mixed solution (C = 4×10^{-5} mol/L; pH 7; for the mixed system SAX: SPX = 4:1).

It is particularly noteworthy that, at the same concentration, the proportion of the long-chain component (SAX) in the mixed system is lower than that in the single-xanthate system. This finding further supports the existence of a synergistic interaction between the long-chain SAX and the short-chain SPX in the mixed system, producing a combined effect greater than the sum of its parts.

The contact angle results are consistent with the flotation test data, explaining the flotation behavior of pure arsenopyrite from the perspective of surface hydrophobicity changes. This alignment also corresponds well with the improved gold recovery observed in the plant following the adoption of the new collector W8.

3.4. Adsorption Amounts of Collector

Measuring the adsorption density of reagents on mineral surfaces can explain why certain collectors can float minerals previously considered refractory, or why specific modifiers render normally floatable minerals difficult to concentrate. Understanding reagent adsorption helps optimize the flotation process, improving both efficiency and recovery. For instance, by adjusting reagent type, concentration, and dosage, the floatability of target minerals can be better controlled, thereby enhancing overall separation performance [29].

Figure 8 shows the relationship between xanthate collector dosage and its adsorption density on arsenopyrite surfaces. The results clearly indicate that under identical concentration conditions, the adsorption density of SAX is higher than that of SPX. This suggests that the stronger collecting power of the longer-chain SAX is attributable not only to a more robust interaction between its molecules and the active sites on the mineral surface but also to a greater number of molecules adsorbed [30].

3. Özçelik, S.; Ekmekçi, Z. Surface Chemistry and Flotation of Gold-Bearing Pyrite. *Minerals* **2024**, *14*, 914. [CrossRef]
4. Gorain, B. FLOT-ART: An Integrated Plant Optimization Tool for the Flotation of Copper and Copper–Gold Ores. *Miner. Eng.* **2024**, *207*, 108578. [CrossRef]
5. Wang, Z.; Zou, D.; Zhao, K.; Chen, R.; Huang, X.; Wang, X. Reducing the Adhesion of Mica on Arsenopyrite Surface Using Sodium Phytate and the Application in Flotation Separation. *Sep. Sci. Technol.* **2022**, *58*, 835–847. [CrossRef]
6. Xing, Q.; Ming, P.; Wang, X.; Li, F.; Wang, Z.; Zhao, K. Fenugreek Polysaccharide Gum as a Depressant in the Flotation Separation of Gold Ore with a High Content of Clay Minerals. *Colloids Interfaces* **2025**, *9*, 21. [CrossRef]
7. Forson, P.; Skinner, W.; Asamoah, R. Investigating the Selective Flotation of Auriferous Arsenian Pyrite from Refractory Ores Using Thionocarbamate. *Powder Technol.* **2023**, *426*, 118649. [CrossRef]
8. Liu, J.; Tao, Y.; Chang, T.; Ge, W.; Jiang, K.; Lv, L.; Zhu, Y.; Yuan, S. Study on Flotation Separation of Barite Fluorite by Citric Acid under New Collector System. *Colloids Surfaces A Physicochem. Eng. Asp.* **2024**, *692*, 134058. [CrossRef]
9. Zhang, W.; Feng, Z.; Xu, S.; Gao, Z. Molecular Design and Performance Evaluation of Collectors for Sulfide Minerals: A Review. *J. Nonferrous Met.* **2025**, *2*, 68–86. [CrossRef]
10. Rao, K.H.; Forssberg, K.S.E. Mixed Collector Systems in Flotation. *Int. J. Miner. Process* **1997**, *51*, 67–79. [CrossRef]
11. Cao, Q.; Cheng, J.; Wen, S.; Li, C.; Bai, S.; Liu, D. A Mixed Collector System for Phosphate Flotation. *Miner. Eng.* **2015**, *78*, 114–121. [CrossRef]
12. Lotter, N.; Bradshaw, D. The Formulation and Use of Mixed Collectors in Sulphide flotation. *Miner. Eng.* **2010**, *23*, 945–951. [CrossRef]
13. Hu, Y.; Wang, J.; Qin, Q.; Cao, Z.; Lu, W.; Shi, J.; Wu, X.; Wang, P.; Sun, Y. Enhancing the Flotation of Columbite via the Synergistic Effect of Mixed Collectors. *Miner. Eng.* **2025**, *225*, 109235. [CrossRef]
14. Wang, Z.; Yan, G.; Yang, L. Experimental Study and Production Practice for Reducing the Tailing Grade in a Copper Smelting Slag Concentrator. *World Nonferrous Met.* **2022**, *31*, 43–47. [CrossRef]
15. Bradshaw, D.J.; Harris, P.J.; O'Connor, C.T. Synergistic Interactions between Reagents in Sulphide flotation. *J. S. Afr. I. Min. Metall.* **1998**, *98*, 189–193. Available online: https://journals.co.za/doi/abs/10.10520/AJA0038223X_2481 (accessed on 15 November 2025).
16. Zhou, X.; Pan, T.; Ding, Q.-F.; Cheng, L.; Song, K.; Liu, F.; Gao, Y. Isotope Geochemistry of the Shenshuitan Gold Deposit within the Wulonggou Gold Field in the Eastern Kunlun Orogen, Northwest China: Implications for Metallogeny. *Minerals* **2022**, *12*, 339. [CrossRef]
17. Zhang, W.; Yuan, Q.; Jia, S.; Li, Z.; Yin, X. Multi-Objective Optimization of Forth Flotation Process: An Application in Gold Ore. *Sustainability* **2021**, *13*, 8314. [CrossRef]
18. Zhao, P.; Liu, W.; Liu, W.; Bao, L.; Shen, Y.; Butt, S.; Zhang, Y. Synthesis, Flotation Behavior, and Structure-Performance Relationship of Three Hydroxyl-Containing Cationic Collectors. *Miner. Eng.* **2025**, *231*, 109443. [CrossRef]
19. Pan, Z.; Sun, X.; Wei, Q.; Jiao, F.; Qin, W. Impact of Serpentine Particle Coating on the Adsorption of Ethyl Xanthate on Pentlandite Surface in Flotation. *Miner. Eng.* **2024**, *216*, 108845. [CrossRef]
20. Zhou, M.; Wang, Z.; Zhao, K.; Yao, C.; Chen, R.; Huang, X.; Safarov, S.S.; Kholov, K.I. Combined Collector Based on N-propyl-N-allyl-o-isobutyl Thiocarbamate Improving the Flotation of Pentlandite. *Miner. Eng.* **2025**, *227*. [CrossRef]
21. Suo, X.; Yu, B.; Sha, J.; Gao, R.; Qi, M.; Huang, Y.; Peng, W.; Guo, X.; Wang, W.; Cao, Y.; et al. Flotation Separation of Quartz and Feldspar under Weak Alkaline Conditions Using Amine ether as a Novel Collector. *Chem. Eng. Sci.* **2025**, *315*, 121873. [CrossRef]
22. Chen, G.; Zhao, H.; Zhou, J.; Liu, Z.; Yang, H. Process Mineralogy Study and Flotation Testwork of a Complex Lead–Gold Rougher Concentrate. *Minerals* **2025**, *15*, 967. [CrossRef]
23. Tamm, K.; Zadeh, Z.A.; Kuusik, R.; Kallas, J.; Yang, J.; Tõnsuaadu, K.; Trikkel, A. Effect of Flotation Time and Collector Dosage on Estonian Phosphorite Beneficiation. *Minerals* **2021**, *11*, 114. [CrossRef]
24. Özün, S.; Ergen, G. Determination of Optimum Parameters for Flotation of Galena: Effect of Chain Length and Chain Structure of Xanthates on Flotation Recovery. *ACS Omega* **2019**, *4*, 1516–1524. [CrossRef] [PubMed]
25. Kwok, D.Y.; Neumann, A.W. Contact Angle Measurement and Contact Angle Interpretation. *Adv. Colloid Interfac.* **1999**, *8*, 167–249. [CrossRef]
26. Kruszelnicki, M.; Polowczyk, I.; Kowalczuk, P.B. Insight into the Influence of Surface Wettability on Flotation Properties of Solid Particles—Critical Contact Angle in Flotation. *Powder Technol.* **2023**, *431*, 119056. [CrossRef]
27. Alghunaim, A.; Kirdponpattara, S.; Newby, B.-M.Z. Techniques for Determining Contact Angle and Wettability of Powders. *Powder Technol.* **2016**, *287*, 201–215. [CrossRef]
28. de Medeiros, A.R.S.; Baltar, C.A.M. Importance of Collector Chain Length in Flotation of Fine Particles. *Min. Eng.* **2018**, *122*, 179–184. [CrossRef]
29. Rao, K.H.; Forssberg, K.S.E. Mechanism of Fatty Acid Adsorption in Salt-type Mineral Flotation. *Min. Eng.* **1991**, *4*, 879–890. [CrossRef]

30. Xie, Y.; Yin, W.; Yao, J.; Yin, X.; Liu, J.; Xue, F.; Tian, D. Flotation Behavior and Surface Adsorption Mechanism of a Novel Selective Inhibitor HDP in the Separation of Chalcopyrite and Talc Flotation. *J. Mol. Liq.* **2024**, *398*, 124206. [CrossRef]
31. Monte, M.B.d.M.; Pimentel, D.A.; Albuquerque, M.D.d.F.d.; Neumann, R.; Silva, L.A.; Correia, J.C.; Uliana, A. Synergism of mixed cationic collectors in the flotation of quartz unveiled by AFM, solution chemistry and quantum chemical calculations. *J. Mol. Liq.* **2023**, *376*. [CrossRef]
32. Zeng, H.; Sun, W.; Tang, H.; Jiang, F.; Wang, L. Surface Roughness and Its Role in Flotation Behavior, Wettability, and Bubble–Particle Interactions: A Systematic Review. *Appl. Sci.* **2025**, *15*, 4557. [CrossRef]
33. Lu, Y.; Liu, D.; Cai, Y.; Gao, C.; Jia, Q.; Zhou, Y. AFM measurement of roughness, adhesive force and wettability in various rank coal samples from Qinshui and Junggar basin, China. *Fuel* **2022**, *317*, 123556. [CrossRef]
34. Zhu, R.; Gu, G.; Chen, Z.; Wang, Y.; Song, S. A new collector for effectively increasing recovery in copper oxide ore-staged flotation. *Minerals* **2019**, *9*, 595. [CrossRef]
35. Şimşek, S.; Kilinç, T.; Cebeci, Y.; Abbaker, A. Application of Flotation Kinetics Models to Chalcopyrite Flotation: Determination of Optimum Flotation Times. *J. Dispers. Sci. Technol.* **2024**, 1–11. [CrossRef]
36. Bahrami, A.; Kazemi, F.; Ghorbani, Y. Effect of Different Reagent Regime on the Kinetic Model and Recovery in Gilsonite Flota-Tion. *J. Mater. Res. Technol.* **2019**, *8*, 4498–4509. [CrossRef]

colloids and interfaces

MDPI

Article

Influence of Sodium Polystyrene Sulfonate on Surface Properties of Dispersions of Oat Globulin Fibrils

Boris A. Noskov [1,*], Alexey G. Bykov [1], Alexandra D. Khrebina [1], Evlaliya A. Levchuk [1], Giuseppe Loglio [2], Reinhard Miller [3] and Egor A. Tsyganov [1]

1 Institute of Chemistry, St. Petersburg State University, 199034 St. Petersburg, Russia; ag-bikov@mail.ru (A.G.B.); sashahrebina27201@gmail.com (A.D.K.); st094543@student.spbu.ru (E.A.T.)

2 Institute of Condensed Matter Chemistry and Technologies for Energy, 16149 Genoa, Italy; giuseppe.loglio@ge.icmate.cnr.it

3 Institute for Condensed Matter Physics, Technical University Darmstadt, 64289 Darmstadt, Germany; reinhard.miller@pkm.tu-darmstadt.de

* Correspondence: b.noskov@spbu.ru

Abstract

The formation of mixed adsorption layers of amyloid fibrils of a plant protein, oat globulin (OG), and a strong polyelectrolyte, sodium polystyrene sulfonate (PSS), at the liquid–gas interface was studied by measurements of the kinetic dependencies of surface tension, dynamic surface elasticity, and ellipsometric angle. The micromorphology of the layers was determined by atomic force microscopy. A strong increase in the surface elasticity was discovered when both components had similar concentrations and formed a network of threadlike aggregates at the interface, thereby explaining the high foam stability in this concentration range. The sequential adsorption of PSS and OG resulted in the formation of thick mixed multilayers and the surface elasticity increased with the number of duplex layers.

Keywords: plant protein adsorption; dynamic surface elasticity; protein—polyelectrolyte interactions; liquid–gas interface; mixed protein—polyelectrolyte multilayers

1. Introduction

The interaction of amyloid fibrils with polyelectrolytes has attracted attention for many years in connection with the problem of neurological disorders. While some polyelectrolytes can promote the growth of proteinaceous deposits in the tissues of patients with amyloid diseases, the other charged polymers can be employed for the destruction of amyloids in the human body [1–4]. More recently, the application of amyloid fibrils for the creation of new materials in various branches of industry and medicine has led to a problem of modulating their properties by various additives, in particular by polyelectrolytes. It has been shown that the addition of charged polysaccharides to the films of fibril-based bioplastics increases their tensile strength and water resistance [5], while the addition of natural polyelectrolytes to amyloid fibril hydrogels and aerogels can improve their microstructure and, thus, their mechanical properties, enhancing, in particular, their resistance to compression [6,7]. Yuan and Solin have recently shown that aerogels containing protein fibrils and an electrically conductive polyelectrolyte can be used as piezoresistive pressure sensors [8]. Delivery systems of bioactive substances are a new application of the complexes of protein fibrils with polyelectrolytes [9,10].

Fibrils in most of the applications of the fibril/polyelectrolyte complexes mentioned above are produced from plant proteins [5,7,9,10]. The exchange of animal-based proteins for their plant-derived counterparts in various technologies is a sustainable development, and it is a consequence of the abundance and the ease of access of plant proteins [11,12]. Moreover, their production is less harmful to the environment compared to animal proteins. At the same time, the intensive investigation of fibrils of plant proteins has only recently started, and their properties, the mechanism of their formation, and the properties of their dispersions remain poorly studied [13,14]. Although it has been shown recently that the surface dilational elasticity of fibril dispersions can exceed the values for native protein solutions [15–17], and protein fibrils can be effective stabilizers of foams and emulsions [18–28], any information on the surface properties of the fibril dispersions is rather scarce and relates mainly to fibrils of animal proteins [16,17,29]. Only a few studies have been devoted to the interactions between fibrils and polyelectrolytes at liquid–fluid interfaces. Peydayesh et al. showed that the interactions of β-lactoglobulin (BLG) with hyaluronic acid led to the formation of an asymmetric and highly ordered structure in the surface layer [30]. Very recently, our group discovered a significant increase in the dilational dynamic surface elasticity of BLG fibril dispersions under the influence of small additions of sodium polystyrene sulfonate (PSS) and studied the formation of mixed multilayers of PSS and BLG fibrils at the water–air surface [31]. In spite of the peculiar properties of the dispersions of plant protein fibrils, their strong interactions with polyelectrolytes in bulk phases [1–7,9,10], and the effective application of these fibrils for the stabilization of foams [11,14,15,27,28], the mixed adsorption layers of plant protein fibrils and strong polyelectrolytes have not been investigated yet, to the best of our knowledge. Therefore, this work is devoted to mixed adsorption layers of a typical synthetic polyelectrolyte, PSS, and fibrils of a plant protein—oat globulin (OG)—at the liquid–air interface. This protein forms typical fibres at elevated temperatures and pH 2, and the properties of their dispersions have been studied by some authors [13,32–34]. One of the aims of this work is to evaluate the peculiarities of the mixed OG/PSS adsorption layers and to determine the conditions required for the development of high dynamic surface elasticity of the mixed dispersions. Another aim consists of the preparation of multilayers at the water–air interface containing fibrils of a plant protein.

2. Materials and Methods

OG was extracted from defatted and ground oat groats according to the procedure described by Zhou et al. [32]. PSS (Mw ≈ 70,000 Da, Sigma-Aldrich, Darmstadt, Germany) was used as received. Protein fibrils were prepared from freeze-dried OG after purification by dialysis. The protein was dissolved in triply distilled water at pH 2 and heated up to 90 °C for 18 h with stirring. After that, the prepared dispersion of mature OG fibrils was centrifuged (12,000× *g*) for 4.5 h to obtain purified fibrils (pFOGs). Protein stock solutions and fibril dispersions were stored in a refrigerator for no longer than 2 days and 1 month, respectively. The solutions of native OG and fibril dispersions of given concentrations were prepared by dilution of the stock solutions and dispersions and mixed a few minutes before measurements at 22 °C. The final fibril concentration of the stock aqueous dispersion was estimated gravimetrically. The fibril concentration of all investigated systems was 20 mg/L, while the PSS concentration varied from 0.2 mg/L to 2 g/L. The pH of the investigated solutions and dispersions was reduced to 3 by HCl additions. This pH was low enough to ensure the solubility of the fibrils in water.

2.1. Surface Tension and Dynamic Surface Elasticity

The surface tension was measured by the Wilhelmy plate method using a ground platinum plate. The accuracy of the measurements was approximately ±0.2 mN/m.

The dynamic dilatational surface elasticity was determined by the oscillating barrier method using the ISR instrument KSV NIMA (Helsinki, Finland), as previously described [29,35]. The oscillations of two barriers in the Langmuir trough with the frequency and amplitude of 0.03 Hz and 4%, respectively, led to oscillations of the surface area and thus of the surface tension.

The real ε_{re} and imaginary ε_{im} components of the complex dilatational dynamic surface elasticity ε were calculated according to the following relation:

$$\varepsilon = \varepsilon_{re} + i \times \varepsilon_{im} = \delta\gamma / \delta lnA,$$

where $\delta\gamma$ and δlnA are the increments (amplitude of oscillations) of the surface tension and relative surface area, respectively. The experimental errors of the oscillating barrier method are mainly determined by the errors of the surface tension measurements and were approximately $\pm 5\%$. The imaginary component of the surface elasticity is significantly smaller than the real component, so only the latter one will be discussed further below.

2.2. Ellipsometry

A null ellipsometer NTEGRA Prima instrument (Optrel-GBR, Berlin, Germany) with a laser wavelength of 623.8 nm was applied to estimate the changes in the surface concentration in the course of adsorption. The ellipsometric angle Δ is expected to be proportional to the surface concentration [36]. The measurements were taken at an angle close to the Brewster angle.

2.3. Atomic Force Microscopy (AFM)

To estimate the micromorphology of the adsorption layer, it was transferred from the liquid surface onto a freshly cleaved mica plate using the Langmuir–Schaeffer method, and the plate with the layer was dried in a desiccator for more than 24 h. After that, the layer on the plate was investigated by the atomic force microscope (NTEGRA Prima instrument, NT-MDT, Moscow, Russia) in a semi-contact mode. The cantilever has an approximate curvature radius of 10 nm.

2.4. Dynamic Light Scattering

The size of the protein particles and their ζ-potential in the bulk phase were determined by dynamic light scattering (DLS) using a Zetasizer ZS Nano analyzer (Malvern Instruments, Malvern, UK). The measurements were carried out at a scattering angle of 173°. The Smoluchowski equation was used in the course of calculating the ζ-potential.

3. Results and Discussion

OG forms relatively rigid fibrils with lengths of up to a few microns, similar to those of BLG and lysozyme fibrils [29], but strongly different from curly fibrils of bovine serum albumin [37]. AFM images of a dispersion drop dried on a mica surface show aggregates of different diameters from 2 up to 6 nm (see Figure S1 of the Supporting Information). The thinnest aggregates are presumably single protofilaments, while the thicker ones consist of a few tightly packed protofilaments.

The dynamic light scattering shows that at low PSS concentrations (<2 mg/L), the size distribution of the aggregates in dispersions of mixed OG/PSS fibrils is bimodal and similar to that of pure OG fibril dispersions (see Figure S2 of the Supporting Information). The main peak of the scattered intensity close to 400 nm presumably corresponds to long fibrils and the smaller peak close to 50 nm can correspond to the admixture of peptides of lower molecular weight. A further increase in PSS concentration leads to a more complex size distribution. The main peak shifts in the direction of larger particle sizes, and at PSS concentrations higher than approximately 5 mg/L, the number of peaks increases, indicating a strongly polydisperse system of mixed aggregates with a possible contribution of free PSS molecules or their aggregates. The largest aggregates corresponding to the main peak at 700 nm appear at a concentration of 20 mg/L.

The increase in the PSS concentration is accompanied by significant changes in the ζ-potential of the aggregates from positive values (~40 mV for pure OG fibrils) to negative ones (~−60 mV at the PSS concentration of 20 mg/L) (Table 1). The change in the sign of the ζ-potential and the subsequent strong increase in its absolute values in this case reflect the effective binding of the negatively charged polyelectrolyte by OG fibrils. The formation of stable large aggregates leads to a decrease in their diffusion coefficients.

Table 1. ζ-potential of the complexes of OG fibrils and PSS.

Concentration Ratio Fibrils/PSS (Fibril Concentration Is Constant and Equals 20 mg/L)	ζ, mV
Only fibrils	36.7
100:1	38.5
10:1	18.9
1:1	−40.8
1:10	−48.2
1:100	−56.6

The growth of the aggregates with high absolute values of ζ-potential leads to an increase in the electrostatic barrier in the course of adsorption and decelerates the formation of adsorption layers.

The kinetic dependencies of surface tension and dynamic surface elasticity of the dispersions of the complexes of OG fibrils and PSS at low PSS concentrations (≤20 mg/L) coincide with the results for pure fibril dispersions without the polyelectrolyte (Figure 1). In this case, the PSS molecules are almost not adsorbed at the water surface due to their low surface activity [38], and the adsorption layer is formed mainly at the expense of the fibril transition from the bulk phase to the surface of the dispersion.

The increase in the PSS concentration leads to significant changes in the adsorption layer structure and surface properties. The formation of complexes of high surface activity from PSS molecules and OG fibrils results in a compaction of the layer structure and an increased dynamic surface elasticity. Noticeable changes in the surface properties occur in the PSS concentration range of 5–100 mg/L, where the steady-state values of the dynamic surface elasticity can be almost two times the values of pure fibril dispersions (Figure 1). The surface elasticity decreases with a further increase in the PSS concentration and reaches almost the values of pure fibril dispersions without any additions. Although PSS starts to decrease the surface tension of water at concentrations higher than approximately 1 g/L, the PSS adsorption is insufficient to explain the changes in the surface properties of the mixed dispersions at PSS concentrations higher than 200 mg/L. Even at the polyelectrolyte concentration of 2 g/L the surface elasticity and the surface pressure of mixed fibrils/PSS dispersions are higher than those of pure PSS solutions (Figures S1 and S3 of the Supporting Information). This means that PSS does not displace fibrils entirely from the interface even if its mass concentration exceeds 100 times that of the fibrils, and some fibrils/PSS complexes are still preserved in the proximal region of the surface layer.

The increase in the PSS concentration from approximately 5 mg/L leads also to the deceleration of the changes in surface properties (Figure 1a,b). This effect can be explained partly by an increase in the electrostatic adsorption barrier when the negative charge of the complexes increases at the expense of the increase in the number of PSS molecules in the complex. In addition, the increase in the size of fibrils/PSS aggregates in the dispersion (cf. Figure S2 of the Supporting Information) is accompanied by a decrease in the diffusion coefficient of the complexes and, thereby, by a deceleration of the mass exchange between the bulk phase and the surface layer. The observed effect is especially significant in the PSS concentration range 5–100 mg/L corresponding to the existence of large fibrils/PSS aggregates.

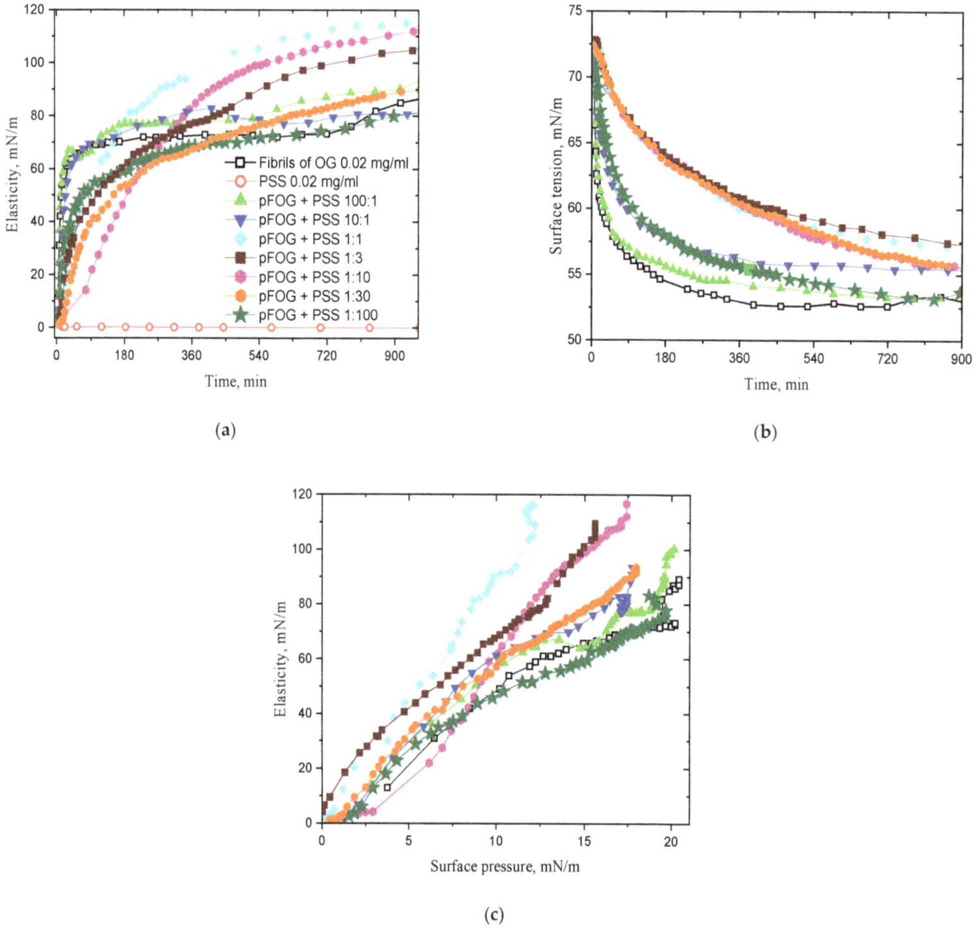

Figure 1. Kinetic dependencies of the dynamic surface elasticity (**a**), dynamic surface tension (**b**); dependencies of the dynamic surface elasticity on surface pressure for OG fibrils/PSS dispersions (**c**) at a fixed OG fibril concentration of 20 mg/L and OG/PSS molar ratios 100:1 (light green triangles), 10:1 (blue triangles), 1:1 (cyan diamonds), 1:3 (wine squares), 1:10 (magenta hexagons), 1:30 (orange circles), and 1:100 (olive asterisks) at pH 3. Open black squares correspond to pure OG dispersions and open red circles to 20 mg/L PSS solutions.

Figure 2 shows the kinetic dependencies of the difference in the ellipsometric angle Δ of the dispersion and its value for pure water Δ_0. This difference decreases with increasing PSS concentration, especially at concentrations above 20 mg/L, and the observed effect corroborates an increase in the polyelectrolyte concentration in the surface layer and the concomitant decrease in the relative fibril surface concentration. Note that the refractive index of the surface layer of PSS solutions is much less than that of the layer of protein solutions at the same concentrations [31]. The adsorbed fibrils/PSS complexes contain a larger amount of PSS molecules at increased polyelectrolyte concentrations and the ellipsometric angle decreases. Simultaneously, the ellipsometric angle increases much more slowly, with the surface age indicating a decrease in the diffusion coefficient of the kinetic units in the bulk phase as a result of aggregate growth.

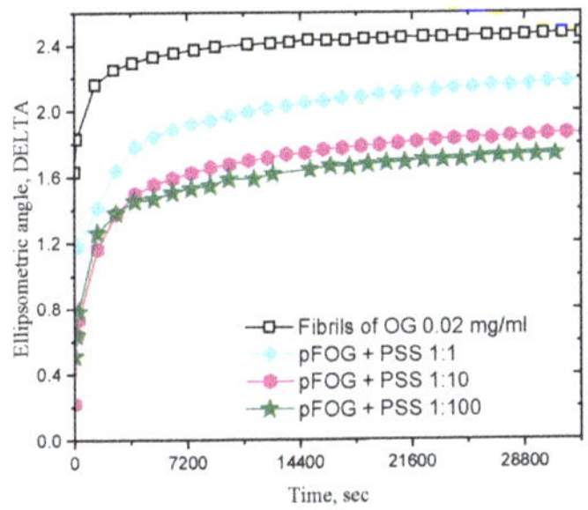

Figure 2. Kinetic dependencies of the ellipsometric angle DELTA for pFOG/PSS adsorption layers at a fixed OG fibril concentration of 20 mg/L and OG/PSS molar ratios 1:1 (cyan diamonds), 1:10 (magenta hexagons), and 1:100 (olive asterisks) at pH 3. Open black squares correspond to pure OG dispersions.

AFM images of the mixed adsorption layers of fibrils/PSS dispersions are similar to those of pure layers of protein fibrils. With the increased PSS concentration, the number of visible separate fibrils in the layer decreases strongly and one can observe mainly a heterogeneous layer without the possibility to distinguish separate aggregates and to estimate their shape (see Figure S4 of the Supporting Information).

The opposite signs of the charges of OG fibrils and PSS molecules at pH 3 allow us to assume that mixed fibrils/PSS multilayers form if the polyelectrolyte molecules and fibrils are adsorbed sequentially at the liquid–gas interface. Multilayers are really formed in the case of a consecutive adsorption of an animal protein, BLG, and the polyelectrolyte PSS [31]. Although PSS is characterized by a relatively weak surface activity, it forms monolayers at the surface of aqueous solutions at a concentration of ~10 g/L. Its adsorption is almost irreversible and the replacement of PSS solution below the adsorption layer by pure water did not lead to noticeable changes in the surface properties. The subsequent replacement of water with a dilute dispersion of BLG fibrils resulted in the formation of a fibril layer at the interface, leading to noticeable changes in the surface properties. This procedure can be repeated a few times, leading to the formation of a relatively thick mixed layer at the interface [31].

Approximately the same procedure was used in this study to prepare a mixed layer of the fibrils of a plant protein, OG, and PSS. The polyelectrolyte adsorption from a 10 g/L solution led to a drop in the surface tension to ~52 mN/m and an increase in the dynamic surface elasticity up to ~45 mN/m (Figure 3a,b). Simultaneously, the ellipsometric angle Δ increased by approximately 1 degree (Figure 4). The subsequent exchange in the PSS solution by water and after that by a 170 mg/L dispersion of OG fibrils resulted in insignificant changes in the surface elasticity and a further drop in the surface tension to ~43 mN/m (Figure 3). The relative changes in the angle Δ were much higher due to the OG fibril adsorption, which was accompanied by a strong increase in the refractive index of the adsorption layer (Figure 4). The subsequent processes of the replacement of the fibril dispersion by PSS solutions and the replacement of these solutions by the fibril dispersions again and so on led to almost periodical changes in the surface properties (Figures 3 and 4). Presumably, this procedure made it possible to obtain a structure of alternating layers of PSS and OG fibrils, as in the case of the mixed heterogeneous adsorption layers of PSS and BLG fibrils showing an alternating layer structure [31] (Figure 5). The surface tension decreases significantly only during the adsorption of the first fibril layer, while the adsorption of the subsequent layers leads only to slighter changes. This peculiarity indicates that the surface tension is determined mainly by the concentration of amphiphilic substances in the first monolayer at the interface. On the contrary, the dynamic surface elasticity shows very little change in the course of the formation of the first three layers, as a result of the relatively loose structure of these layers, unlike stronger changes during the formation of the subsequent denser duplex layers [31,39]. These findings are in agreement with the results of the preceding studies on the dependence of the surface elasticity on the thickness of the interfacial layer [39–41] and the inhomogeneity of the first PSS adsorption layer [38]. Unlike the dynamic surface elasticity and surface tension, the ellipsometric angle is an approximately linear function of the number of duplex layers (Figure 4).

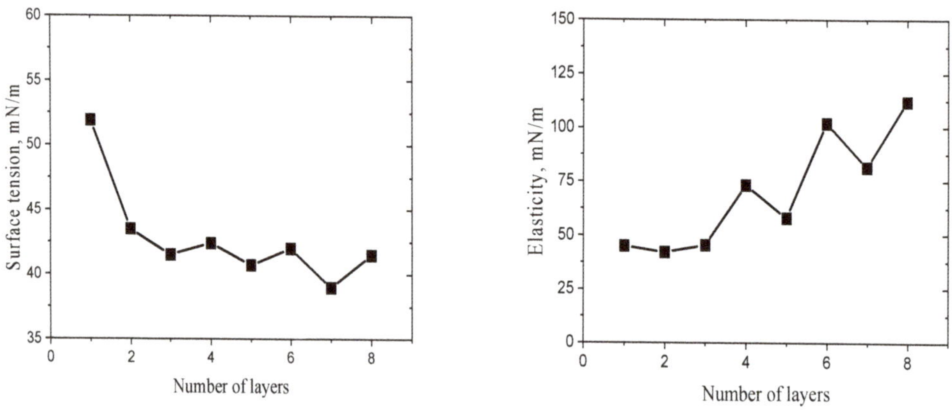

(**a**) (**b**)

Figure 3. Dependencies of surface tension (**a**) and dynamic surface elasticity (**b**) as a function of the number of monolayers of OG fibrils and PSS in the adsorption multilayer. The pH of subphase was 3.

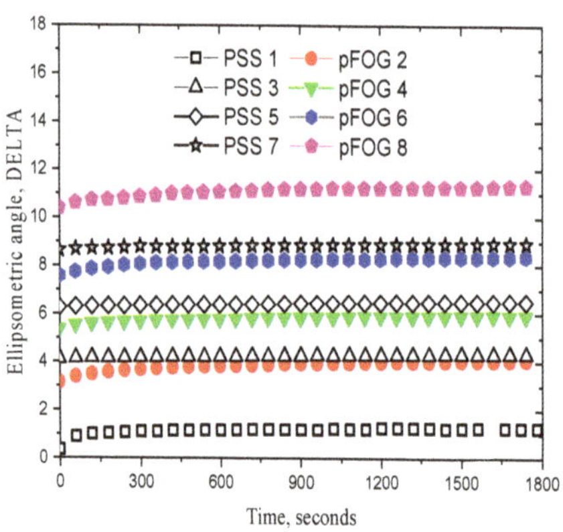

Figure 4. Kinetic dependencies of the ellipsometric angle Δ of mixed multilayers formed by the consecutive adsorption of PSS (black open symbols) and OG fibrils (closed coloured symbols): 1 (squares), 2 (circles), 3 (triangles up), 4 (triangles down), 5 (diamonds), 6 (hexagons), 7 (asterisks), and 8 (pentagons) layers. The pH of the subphase was 3.

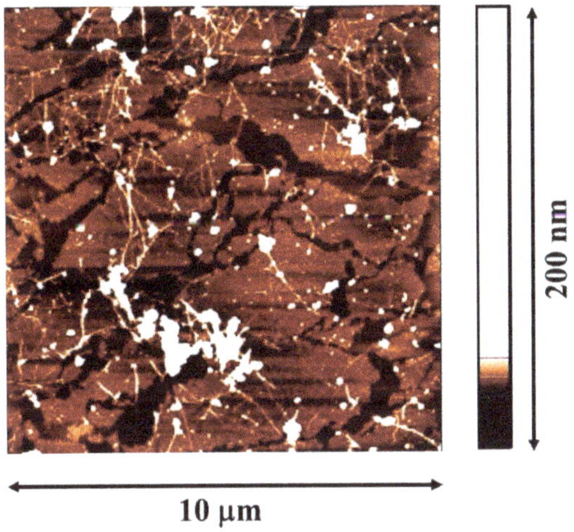

Figure 5. The AFM image of 8 OG fibrils/PSS layers. The multilayer was transferred from the water–air surface on the surface of mica.

The obtained results show that the fibrils of plant proteins can also form thick mixed layers with strong polyelectrolytes at liquid–fluid interfaces, like the fibrils of animal proteins [31]. There are only a few not very significant quantitative differences between the dependencies of the surface properties of the multilayers on the number of adsorption cycles for OG fibrils/PSS and BLG fibrils/PSS systems. In the latter case, for example, the dynamic surface elasticity is somewhat higher, ~185 mN/m against 115 mN/m in the latter case. This difference in the surface elasticity can lead to the higher stability of foams for BLG fibrils/PSS dispersions, but it is difficult to anticipate a significant difference in the foam stability because both values are high—much higher than the surface elasticity of the corresponding native protein solutions.

4. Conclusions

The influence of a polyelectrolyte on the surface properties of dispersions of plant protein fibrils has been studied for the first time, to the best of our knowledge. Although the surface activity of PSS is low, it starts to change the surface properties of the dispersions of OG fibrils at concentrations much less than those corresponding to a noticeable decrease in the surface tension of water, thereby indicating the formation of fibrils/PSS complexes in the surface layer. If the mass concentrations of OG fibrils and PSS are comparable, the surface properties change significantly and the dynamic surface elasticity of the mixed dispersions can exceed twice the values of pure fibril dispersions. This effect can indicate a strong impact of polyelectrolytes on the properties of foams and emulsions stabilized by plant proteins. A further increase in PSS concentration does not lead to the complete displacement of the protein from the interface, and the complexes are still present in the surface layer. The consecutive adsorption of protein and polyelectrolyte at the dispersion–air interface results in the formation of thick fibrils/PSS multilayers with a high dynamic surface elasticity (>100 mN/m), if the number of cycles of fibril and polyelectrolyte adsorption is sufficiently high. Although these values are somewhat lower than in the case of multilayers of an animal protein (BLG) and PSS, they can result in peculiar properties of the corresponding dispersion systems. If the number of adsorbed layers is less than about three, the dynamic elasticity is almost close to that of a monolayer, presumably due to a loose and heterogeneous surface structure in this case.

Supplementary Materials: The following Supporting Information can be downloaded at https://www.mdpi.com/article/10.3390/colloids9060089/s1, Figure S1: AFM image of OG unpurified fibrils; Figure S2: DLS results for the dispersions of OG fibrils (A) and OG fibrils/PSS at an OG/PSS molar ratio of 10:1 (B); Figure S3: Kinetic dependencies of the dynamic surface elasticity (A), dynamic surface tension (B); dependencies of the dynamic surface elasticity on surface pressure for PSS solutions (C). Filled black squares correspond to 20 mg/L PSS solutions, open magenta circles to 2 g/L PSS solutions; Figure S4: AFM images of mixed adsorption layers of OG fibrils/PSS complexes at various OG/PSS molar ratios: 100:1 (A), 10:1 (B), 1:1 (C), 1:10 (D).

Author Contributions: Conceptualization, A.G.B. and B.A.N.; methodology, A.G.B.; software, G.L.; validation, E.A.T. and R.M.; formal analysis, E.A.T.; investigation, E.A.L. and A.G.B.; data curation, A.D.K. and R.M.; writing—original draft preparation, B.A.N. and A.D.K.; writing—review and editing, R.M., G.L. and B.A.N.; supervision, B.A.N., A.G.B. and E.A.L. All authors have read and agreed to the published version of the manuscript.

Funding: The authors acknowledge Saint-Petersburg State University for providing funding (research project Pure ID: 131065423).

Data Availability Statement: The data presented in this study are available on request from the corresponding author.

Acknowledgments: The use of the equipment of the Resource Centre for Diagnostics of Functional Materials for Medicine, Pharmacology and Nanoelectronics; the Chemical Analysis and Materials Research Centre; the Centre for Diagnostics of Functional Materials for Medicine, Pharmacology and Nanoelectronics; the Interdisciplinary Resource Centre for Nanotechnology; and the Resource Centre for Molecular and Cell Technologies of SPbU is gratefully acknowledged.

Conflicts of Interest: The authors declare no conflicts of interest.

References

1. Calamai, M.; Kumita, J.R.; Mifsud, J.; Parrini, C.; Ramazzotti, M.; Ramponi, G.; Taddei, N.; Chiti, F.; Dobson, C.M. Nature and Significance of the Interactions between Amyloid Fibrils and Biological Polyelectrolytes. *Biochemistry* **2006**, *45*, 12806–12815. [CrossRef]
2. Semenyuk, P.; Kurochkina, L.; Barinova, K.; Muronetz, V. Alpha-Synuclein Amyloid Aggregation Is Inhibited by Sulfated Aromatic Polymers and Pyridinium Polycation. *Polymers* **2020**, *12*, 517. [CrossRef]
3. Islam, M.; Argueta, E.; Wojcikiewicz, E.P.; Du, D. Effects of Charged Polyelectrolytes on Amyloid Fibril Formation of a Tau Fragment. *ACS Chem. Neurosci.* **2022**, *13*, 3034–3043. [CrossRef]
4. Makshakova, O.; Bogdanova, L.; Faizullin, D.; Khaibrakhmanova, D.; Ziganshina, S.; Ermakova, E.; Zuev, Y.; Sedov, I. The Ability of Some Polysaccharides to Disaggregate Lysozyme Amyloid Fibrils and Renature the Protein. *Pharmaceutics* **2023**, *15*, 624. [CrossRef]
5. Zhang, Y.; Nian, Y.; Shi, Q.; Hu, B. Protein Fibrillation and Hybridization with Polysaccharides Enhance Strength, Toughness, and Gas Selectivity of Bioplastic Packaging. *J. Mater. Chem. A* **2023**, *11*, 9884–9901. [CrossRef]
6. Usuelli, M.; Germerdonk, T.; Cao, Y.; Peydayesh, M.; Bagnani, M.; Handschin, S.; Nyström, G.; Mezzenga, R. Polysaccharide-Reinforced Amyloid Fibril Hydrogels and Aerogels. *Nanoscale* **2021**, *13*, 12534–12545. [CrossRef]
7. Chen, Q.-H.; Li, X.-Y.; Huang, C.-L.; Liu, P.; Zeng, Q.-Z.; Yang, X.-Q.; Yuan, Y. Development and Mechanical Properties of Soy Protein Isolate-Chitin Nanofibers Complex Gel: The Role of High-Pressure Homogenization. *Lebensm.-Wiss. Technol.* **2021**, *150*, 112090. [CrossRef]
8. Yuan, Y.; Solin, N. Protein-Based Flexible Conductive Aerogels for Piezoresistive Pressure Sensors. *ACS Appl. Bio Mater.* **2022**, *5*, 3360–3370. [CrossRef] [PubMed]
9. Yang, Q.; Yu, X.-H.; Zhang, J.-W.; Jiang, Y.-X.; Chen, H.-Q. The Complex of Arachin Amyloid-like Fibrils Formed with Ultrasound Treatment and Chitosan: A Potential Vehicle for Betanin and Curcumin with Improved Chemical Stability and Slow Release In Vitro. *Food Hydrocoll.* **2025**, *166*, 111272. [CrossRef]
10. Chen, H.; Zhou, M.; Xu, Z.; Dong, X.; Ding, X.; Zhou, X.; Cui, P. Novel Fava Bean 11S Nanofiber Gels for Sustained Ergothioneine Delivery: A Calcium Ion and κ-Carrageenan Approach. *Food Hydrocoll.* **2025**, *169*, 111604. [CrossRef]
11. Li, T.; Zhou, J.; Peydayesh, M.; Yao, Y.; Bagnani, M.; Kutzli, I.; Chen, Z.; Wang, L.; Mezzenga, R. Plant Protein Amyloid Fibrils for Multifunctional Sustainable Materials. *Adv. Sustain. Syst.* **2023**, *7*, 2200414. [CrossRef]

12. Peydayesh, M.; Bagnani, M.; Soon, W.L.; Mezzenga, R. Turning Food Protein Waste into Sustainable Technologies. *Chem. Rev.* **2023**, *123*, 2112–2154. [CrossRef] [PubMed]
13. Xu, J.; Tang, M.; Wang, D.; Xie, Q.; Xu, X. Exploring the Self-Assembly Journey of Oat Globulin Fibrils: From Structural Evolution to Modified Functionality. *Food Hydrocoll.* **2024**, *149*, 109587. [CrossRef]
14. Liang, Y.; Zhang, P.; Liu, M.; Liu, H.; He, B.; Zhu, Y.; Wang, J. Plant-Based Protein Amyloid Fibrils: Origins, Formation, Extraction, Applications, and Safety. *Food Chem.* **2025**, *469*, 142559. [CrossRef] [PubMed]
15. Wan, Z.; Yang, X.; Sagis, L.M.C. Nonlinear Surface Dilatational Rheology and Foaming Behavior of Protein and Protein Fibrillar Aggregates in the Presence of Natural Surfactant. *Langmuir* **2016**, *32*, 3679–3690. [CrossRef] [PubMed]
16. Noskov, B.; Loglio, G.; Miller, R.; Milyaeva, O.; Panaeva, M.; Bykov, A. Dynamic Surface Properties of α-Lactalbumin Fibril Dispersions. *Polymers* **2023**, *15*, 3970. [CrossRef]
17. Milyaeva, O.Y.; Akentiev, A.V.; Bykov, A.G.; Loglio, G.; Miller, R.; Portnaya, I.; Rafikova, A.R.; Noskov, B.A. Dynamic Properties of Adsorption Layers of κ-Casein Fibrils. *Langmuir* **2023**, *39*, 15268–15274. [CrossRef]
18. Oboroceanu, D.; Wang, L.; Magner, E.; Auty, M.A.E. Fibrillization of Whey Proteins Improves Foaming Capacity and Foam Stability at Low Protein Concentrations. *J. Food Eng.* **2014**, *121*, 102–111. [CrossRef]
19. Peng, J.; Simon, J.R.; Venema, P.; van der Linden, E. Protein Fibrils Induce Emulsion Stabilization. *Langmuir* **2016**, *32*, 2164–2174. [CrossRef]
20. Loveday, S.M.; Anema, S.G.; Singh, H. β-Lactoglobulin Nanofibrils: The Long and the Short of it. *Int. Dairy J.* **2017**, *67*, 35–45. [CrossRef]
21. Peng, D.; Yang, J.; Li, J.; Tang, C.; Li, B. Foams Stabilized by β-Lactoglobulin Amyloid Fibrils: Effect of PH. *J. Agric. Food Chem.* **2017**, *65*, 10658–10665. [CrossRef]
22. Mantovani, R.A.; de Figueiredo Furtado, G.; Netto, F.M.; Cunha, R.L. Assessing the Potential of Whey Protein Fibril as Emulsifier. *J. Food Eng.* **2018**, *223*, 99–108. [CrossRef]
23. Hu, J.; Yang, J.; Xu, Y.; Zhang, K.; Nishinari, K.; Phillips, G.O.; Fang, Y. Comparative Study on Foaming and Emulsifying Properties of Different Beta-Lactoglobulin Aggregates. *Food Funct.* **2019**, *10*, 5922–5930. [CrossRef]
24. Jiang, F.; Pan, Y.; Peng, D.; Huang, W.; Shen, W.; Jin, W.; Huang, Q. Tunable Self-Assemblies of Whey Protein Isolate Fibrils for Pickering Emulsions Structure Regulation. *Food Hydrocoll.* **2022**, *124*, 107264. [CrossRef]
25. Han, Y.; Zhu, L.; Karrar, E.; Qi, X.; Zhang, H.; Wu, G. Pickering Foams Stabilized by Protein-Based Particles: A Review of Characterization, Stabilization, and Application. *Trends Food Sci. Technol.* **2023**, *133*, 148–159. [CrossRef]
26. Huyst, A.M.R.; Van der Meeren, P.; Housmans, J.A.J.; Monge-Morera, M.; Rousseau, F.; Schymkowitz, J.; Delcour, J.A. Improved Coalescence and Creaming Stability of Structured Oil-in-Water Emulsions and Emulsion Gels Containing Ovalbumin Amyloid-like Fibrils Produced by Heat and Enzymatic Treatments. *Food Hydrocoll.* **2023**, *145*, 109142. [CrossRef]
27. Su, R.; Zhang, X.; Cao, J.; Xu, T.; Liu, S.; Zhang, H.; Wu, D.; Wang, Z.; Tong, X.; Wang, H.; et al. Exploring the Interfacial Behavior and Foam Characteristics of Various Soy Protein Aggregates: Insights of Morphology and Conformational Flexibility. *Food Hydrocoll.* **2025**, *166*, 111362. [CrossRef]
28. Chen, L.; Zhang, C.; Yu, X.; Fu, L.; Tang, X.; Feng, X. Enhance Quinoa Protein Foaming Properties through Amyloid-like Fibrillation and 2S Albumin Blending. *Food Hydrocoll.* **2025**, *164*, 111154. [CrossRef]
29. Noskov, B.A.; Akentiev, A.V.; Bykov, A.G.; Loglio, G.; Miller, R.; Milyaeva, O.Y. Spread and Adsorbed Layers of Protein Fibrils at Water –Air Interface. *Colloids Surf. B Biointerfaces* **2022**, *220*, 112942. [CrossRef]
30. Peydayesh, M.; Kistler, S.; Zhou, J.; Lutz-Bueno, V.; Victorelli, F.D.; Meneguin, A.B.; Spósito, L.; Bauab, T.M.; Chorilli, M.; Mezzenga, R. Amyloid-Polysaccharide Interfacial Coacervates as Therapeutic Materials. *Nat. Commun.* **2023**, *14*, 1848. [CrossRef] [PubMed]
31. Bykov, A.G.; Loglio, G.; Miller, R.; Tsyganov, E.A.; Wan, Z.; Noskov, B.A. Mixed Adsorption Mono- and Multilayers of ß-Lactoglobulin Fibrils and Sodium Polystyrene Sulfonate. *Colloids Interfaces* **2024**, *8*, 61. [CrossRef]
32. Zhou, J.; Li, T.; Peydayesh, M.; Usuelli, M.; Lutz-Bueno, V.; Teng, J.; Wang, L.; Mezzenga, R. Oat Plant Amyloids for Sustainable Functional Materials. *Adv. Sci.* **2022**, *9*, 2104445. [CrossRef] [PubMed]
33. Khrebina, A.D.; Akentiev, A.V.; Wan, Z.; Noskov, B.A. Dynamic Surface Properties of Oat Protein Dispersions. *Mendeleev Commun.* **2025**, *35*, 202–204. [CrossRef]
34. Xu, J.; Tang, M.; Xu, X. Effect of Ultrasound Pretreatment on the Fibrillization of Oat Globulins: Aggregation Kinetics, Structural Evolution, and Core Composition. *Food Hydrocoll.* **2025**, *165*, 111233. [CrossRef]
35. Noskov, B.A.; Krycki, M.M. Formation of Protein/Surfactant Adsorption Layer as Studied by Dilational Surface Rheology. *Adv. Colloid Interface Sci.* **2017**, *247*, 81–99. [CrossRef]
36. Motschmann, H.; Teppner, R. Ellipsometry in Interface Science. In *Novel Methods to Study Interfacial Layers*; Elsevier: Amsterdam, The Netherlands, 2001; Volume 11, pp. 2–42.
37. Akentiev, A.; Lin, S.-Y.; Loglio, G.; Miller, R.; Noskov, B. Surface Properties of Aqueous Dispersions of Bovine Serum Albumin Fibrils. *Colloids Interfaces* **2023**, *7*, 59. [CrossRef]

38. Noskov, B.A.; Nuzhnov, S.N.; Loglio, G.; Miller, R. Dynamic Surface Properties of Sodium Poly(Styrenesulfonate) Solutions. *Macromolecules* **2004**, *37*, 2519–2526. [CrossRef]
39. Pivard, S.; Jacomine, L.; Kratz, F.S.; Foussat, C.; Lamps, J.-P.; Legros, M.; Boulmedais, F.; Kierfeld, J.; Schosseler, F.; Drenckhan, W. Interfacial Rheology of Linearly Growing Polyelectrolyte Multilayers at the Water–Air Interface: From Liquid to Solid Viscoelasticity. *Soft Matter* **2024**, *20*, 1347–1360. [CrossRef] [PubMed]
40. Safouane, M.; Miller, R.; Möhwald, H. Surface Viscoelastic Properties of Floating Polyelectrolyte Multilayers Films: A Capillary Wave Study. *J. Colloid Interface Sci.* **2005**, *292*, 86–92. [CrossRef]
41. Cramer, A.D.; Dong, W.-F.; Benbow, N.L.; Webber, J.L.; Krasowska, M.; Beattie, D.A.; Ferri, J.K. The Influence of Polyanion Molecular Weight on Polyelectrolyte Multilayers at Surfaces: Elasticity and Susceptibility to Saloplasticity of Strongly Dissociated Synthetic Polymers at Fluid–Fluid Interfaces. *Phys. Chem. Chem. Phys.* **2017**, *19*, 23781–23789. [CrossRef]

colloids and interfaces

MDPI

Article

Wetting Behavior of Cationic and Anionic Surfactants on Hydrophobic Surfaces: Surface Tension and Contact Angle Measurements

Sujit Kumar Shah, Rojina Bhattarai, Sujata Gautam, Pawan Shah and Ajaya Bhattarai *

Department of Chemistry, Mahendra Morang Adarsh Multiple Campus, Tribhuvan University, Biratnagar 56613, Nepal; sujit.shah@mmamc.tu.edu.np (S.K.S.); pawan.795703@iost.tu.edu.np (P.S.)
* Correspondence: ajaya.bhattarai@mmamc.tu.edu.np

Abstract

In this study, cationic surfactant cetyltrimethylammonium bromide (CTAB) and anionic surfactant sodium bis(2-ethylhexyl) sulfosuccinate (AOT) are employed to systematically investigate surface and wetting properties on hydrophobic surfaces, specifically in mixed solvents composed of ethylene glycol (EG) and water at 298.15 K. By varying the concentration of each surfactant within the EG–water mixture, both surface tension and contact angle measurements are performed to elucidate how surfactant type and solvent composition influence interfacial behavior and wettability. PTFE and wax surfaces were chosen as model hydrophobic surfaces. Surface tension measurements obtained in pure water and in water–EG mixtures containing 5, 10, and 20 volume percentage EG reveal a consistent decrease in the premicellar slope ($\frac{d\gamma}{dlogC}$) with increasing EG content. This reduction reflects weakened hydrophobic interactions and less effective surfactant adsorption at the air–solution interface. The corresponding decline in maximum surface excess (Γ_{max}) and increase in minimum area per molecule (A_{min}) confirm looser interfacial packing due to EG participation in the solvation layer. Plots of adhesion tension (A_T) versus surface tension (γ) exhibit negative slopes, consistent with reduced solid–liquid interfacial tension (Γ_{LG}) and greater redistribution of surfactant molecules toward the solid–liquid interface. AOT shows stronger sensitivity to EG compared to CTAB, reflecting structural headgroup-specific adsorption behavior. Work of adhesion (W_A) measurements demonstrate enhanced wettability at higher EG concentrations, highlighting the cooperative impact of co-solvent environment and surfactant type on wetting phenomena.

Keywords: ethylene glycol (EG); adhesion tension; work of adhesion; PTFE; wax

1. Introduction

Wetting is a fundamental interfacial phenomenon describing the interaction of liquids with solids, and it plays a crucial role in diverse scientific and technological processes such as coating, lubrication, detergency, inkjet printing, and drug delivery [1]. The degree of wetting is generally characterized by the contact angle (CA), defined as the angle between the liquid–vapor interface and the solid surface at the three–phase contact line [2,3]. According to Young's equation, the equilibrium CA is determined by the balance of interfacial tensions at the solid–liquid, solid–vapor, and liquid–vapor interfaces. Hydrophilic surfaces typically exhibit low contact and angles (<90°), whereas hydrophobic surfaces display high contact angles (>90°), and superhydrophobic surfaces display high values exceeding 150°, indicating extreme repellency [4].

and spreading behavior. By analyzing the interplay between surfactant type, co-solvent presence, and surface characteristics, this research seeks to provide a comprehensive understanding of the factors that govern wetting and interfacial phenomena on low-energy surfaces.

2. Experimental Section

2.1. Materials

CTAB (99%) and AOT (99%) were purchased from Loba Chemi, Mumbai, India, and dried in an oven at 50 °C for 30 min before use. EG (>99%) was purchased from Thermo Fischer Scientific, Mumbai, India. Dental wax sheets (commercial dental-grade wax) used in this study were obtained from Pyrax Polymers, located on Sunhera Road, Roorkee, India. Polytetrafluoroethylene (PTFE) films were purchased as commercial Teflon sheets with a thickness of 0.1 mm from Fluoroplast Engineers, Mumbai, India. For wettability experiments, the PTFE sheets were carefully cut into samples measuring 2 cm × 2 cm.

2.2. Preparation of EG–Water–Surfactant Solutions

All solutions were prepared using double-distilled water (specific conductance less than 2 μScm^{-1}). Mixed solvent systems were obtained by blending water with different volume fractions of EG. Mixed solvent systems of EG and water were prepared by mixing double-distilled water with EG at 5%, 10%, and 20% by volume at 298.15 ± 0.5 K. The concentration ranges of CTAB and AOT solutions were 1×10^{-2} M to 5×10^{-5} M and 3×10^{-2} M to 5×10^{-5} M, respectively. All the solutions were equilibrated at 298.15 ± 0.5 K for at least 24 h prior to measurements to ensure thermodynamic equilibrium.

2.3. Contact Angle and Surface Tension Measurements

The static contact angle of surfactant solutions on hydrophobic surfaces was measured using a Drop Shape Analyzer (DSA, 25E, Kruss, Hamburg, Germany) equipped with a high-resolution camera and Kruss advanced software (version 1.9.0.8). Aqueous solutions of CTAB and AOT surfactants were prepared in water and in ethylene glycol (EG)–water mixed solvents containing 5%, 10%, and 20% of EG by volume. Contact angle measurements were performed using the sessile drop method, which is a widely accepted technique for wettability characterization.

Prior to each measurement, the syringe was calibrated, and a stainless-steel needle with a diameter of 0.5 mm was used for drop formation. The syringe was thoroughly cleaned by rinsing with first and second distilled water, followed by drying with acetone to remove residual moisture. A droplet of known volume (2 µL) was gently deposited onto the substrate surface using a precision microsyringe. The droplet profile was analyzed by fitting the Young–Laplace equation through the instrument software [3,7].

The already cut pieces of dental wax substrate and PTFE films were placed on the sample holder. The surfaces were kept as smooth as possible. A baseline was manually created on the solid surface using the curved baseline option in the software, and the contact angle was determined by selecting the appropriate area of interest (as shown in Scheme 1). For each droplet, multiple frames were recorded, and 20–30 readings were obtained to ensure stable fitting. Each measurement was repeated at three different locations on the same surface, and the average contact angle values were taken for further calculations. All the experiments were conducted at 298.15 ± 0.5 K under a controlled temperature using a circulatory water bath procured from Orbit Pvt. Ltd., Hyderabad, India.

https://doi.org/10.3390/colloids10010008

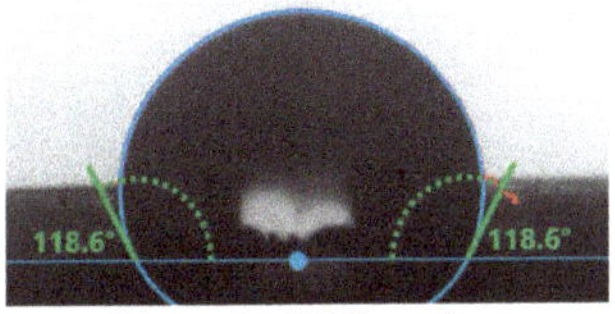

(**a**) On the PTFE surface

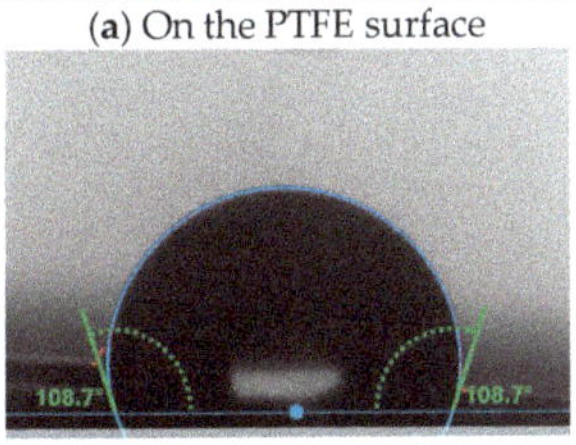

(**b**) On the wax surface

Scheme 1. Images of water drop showing contact angles; the value of contact angle on PTFE surface. The value we obtained is closely related to the literature value 117.4° [2].

Surface tension of the surfactant solutions was measured using the same DSA, 25E, via the pendant drop method. A droplet solution was suspended from the tip of a microsyringe, and its profile was recorded with a resolution camera. The surface tension was obtained by fitting the droplet shape to the Young–Laplace equation using the instrument software. All measurements were performed in triplicate to maintain accuracy and confirm reproducibility.

3. Results and Discussion

3.1. Surface Tension and Surface Properties

Surface tension (γ) measurements were carried out for both CTAB and AOT in pure water as well as in water–EG mixed solvents containing 5%, 10%, and 20% EG by volume at 298.15 K. The data were plotted as surface tension versus the logarithm of surfactant concentration to observe the relationship between concentration and surface tension. As anticipated, increasing the concentration of surfactant led to a progressive decrease in surface tension, which eventually plateaued at the critical micelle concentration (CMC), where additional surfactant molecules aggregate into micelles rather than further reducing surface tension. This reduction in surface tension is primarily due to the adsorption of surfactant molecules at the air–solution interface, effectively lowering the surface free energy. Eventually, the surface tension shows a break point—this point is called the critical micelle concentration (CMC), as shown in Figure 1. At CMC, the surface becomes saturated with surfactant molecules, and the further addition of surfactant forms micelles the in solution rather than affecting surface tension.

The premicellar slope of the graph, $\frac{d\gamma}{dlogC}$, represents the pattern of surface tension decrease with the $logC$. This slope is useful in determining surface properties such as surface excess concentration Γ_{max} using Equation (1):

$$\Gamma_{max} = -\frac{1}{2.303nRT}\left[\frac{d\gamma}{dlogC}\right]_{T,P} \quad (1)$$

https://doi.org/10.3390/colloids10010008

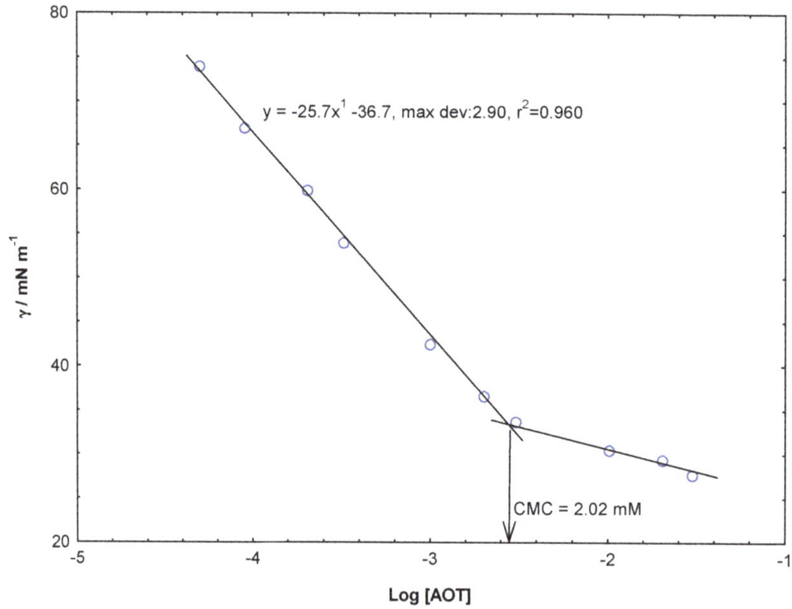

Figure 1. Variation in surface tension with concentration of AOT in water showing CMC at 298.15 K.

R (8.314 J mol^{-1} K^{-1}) denotes the universal gas constant, T indicates the absolute temperature, and C indicates the concentration of surfactant. The value of n taken as the constant is a pre-factor that is taken as 2 for the normal surfactant.

Minimum area per molecule, A_{min}, represents the smallest surface area occupied by a molecule at the air–solution interface. It is determined by using Equation (2)

$$A_{min} = 1/N_A \Gamma_{max} \quad (2)$$

where N_A denotes Avogadro's Number.

Table 1 lists the CMC $\frac{d\gamma}{dlogC}$, Γ_{max}, and A_{min} of AOT and CTAB in water, and the various volume percentages of EG at 298.15 K.

Table 1. Surface properties in water and different volume % of EG at 298.15 K.

Volume % of EG	CMC (mM)	$\left(\frac{d\gamma}{dlogC}\right)$	$\Gamma_{max}10^6$ (molm^{-2})	$A_{min}(\mathring{A}^2$molecule$^{-1})$
AOT				
0	2.02	−16.40	1.44	115.61
5%	3.06	−15.33	1.34	123.68
10%	4.23	−11.23	0.98	168.83
20%	6.12	−10.23	0.90	185.33
CTAB				
0	0.98	−31.00	2.72	61.16
5%	1.12	−27.00	2.36	70.22
10%	1.81	−23.20	2.03	81.72
20%	2.23	−20.80	1.82	91.15

Ethylene glycol (EG) is a strongly hydrogen-bonding, moderately polar organic co-solvent with a dielectric constant of 37 at 25 °C. When mixed with water, it breaks part of the three-dimensional H-bond network and reduces water's cohesive forces, producing a

https://doi.org/10.3390/colloids10010008

monotonic fall in the solvent's surface tension. This trend has been seen in Figures 2 and 3, where an increase in the volume % of EG causes a decrease in the surface tension of both the surfactants.

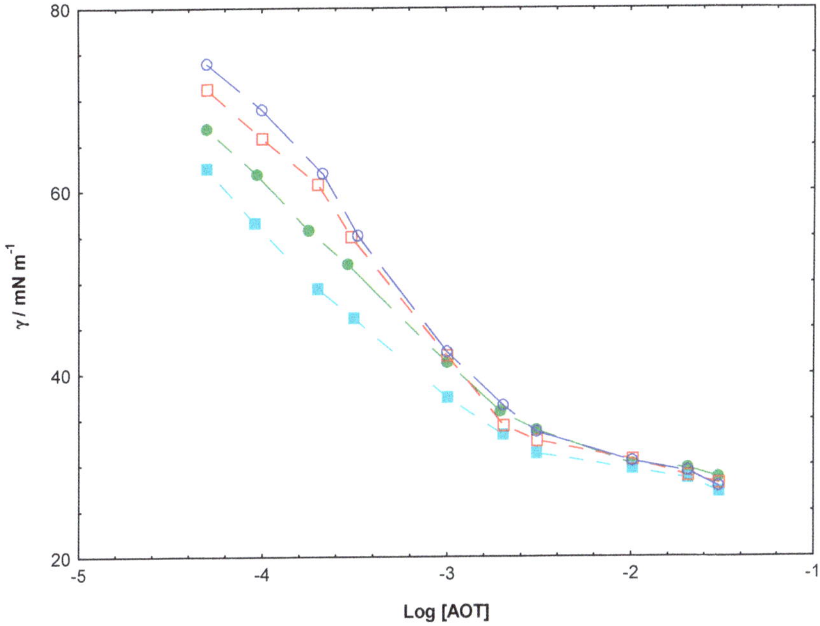

Figure 2. Variation in surface tension with logarithm of concentration of AOT in water (○) and different volume % (5% (□), 10% (•), and 20% (■)) of EG at 298.15 K.

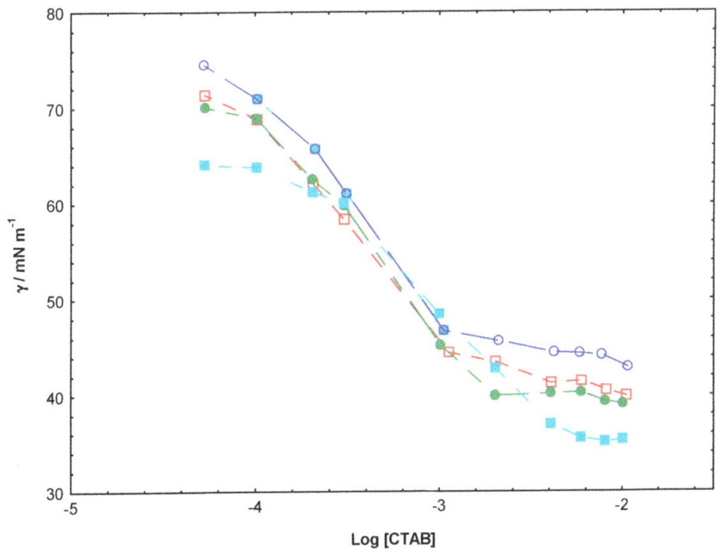

Figure 3. Variation in surface tension with logarithm of concentration of CTAB in water (○) and different volume % (5% (□), 10% (•), and 20% (■)) of EG at 298.15 K.

The CMC of both surfactants increases systematically with EG content (Table 1), although the magnitude of the shift differs for the two ionic types. In water, CTAB exhibits

https://doi.org/10.3390/colloids10010008

0.98 mM, whereas AOT's CMC is 2.02 mM. The upward shift in CMC with increasing EG content is attributed to weakened hydrophobic interactions and reduced dielectric screening in the solvent. The addition of EG lowers the solvent polarity, thereby diminishing the contrast between the hydrophobic tail and solvent environment, which makes micelle formation less favorable. Lower dielectric constant also increases Coulombic repulsion between ionic headgroups, further destabilizing the micelle for both cationic and anionic surfactants [22].

The premicellar slope $\left(\frac{d\gamma}{dlogC}\right)$ of the plot of the surface tension vs. logarithm of molar concentration plot is characterized by a steep linear decrease in surface tension, governed by the progressive accumulation of surfactant molecules at the air–solution interface. Figures 1–3 show that the premicellar slope becomes less steep as the EG volume % increases. The numerical values are presented in Table 1. The reduction in the premicellar slope arises from two concurrent mechanisms. The first one is that EG molecules themselves adsorb at the air–solution interface, and the second is that the dielectric constant of EG–water mixtures is much lower than that of water. This decreases the strength of hydrophobic interactions that drive surfactant adsorption and increases the electrostatic repulsion between ionic headgroups.

The presence of EG in aqueous medium significantly influences the surface excess concentration (Γ_{max}) of both CTAB and AOT, as presented in Table 1. In pure water, the high polarity and extensive H-bond network of water provide a strong driving force for surfactant adsorption at the air–solution interface. As the surfactant concentration increases, molecules rapidly accumulate at the interface, displacing water molecules and reducing surface free energy. Consequently, both surfactants exhibit their highest Γ_{max} values in water. When EG is introduced into the solvent mixture, Γ_{max} decreases systematically with increasing EG volume %. This decline confirms that fewer surfactant molecules occupy a unit interfacial area in the EG-rich environment.

The corresponding minimum area per molecule (A_{min}) shows the inverse trend expanding as EG volume % increases. This expansion signifies looser molecular packing at the interface, consistent with decreased Γ_{max}. The increase in (A_{min}) arises primarily because EG molecules intercalate between surfactant headgroups, diluting the interfacial film and causing a lateral spread of surfactant molecules. The weaker hydrophobic contrast between surfactant tails and the EG–water subphase also reduces tail–tail interaction, so the molecules adopt a more tilted or disordered orientation at the interface, further enlarging the effective A_{min}.

Comparing the two surfactants, CTAB consistently exhibits higher Γ_{max} and smaller A_{min} values than AOT in all solvent compositions, demonstrating more efficient packing and stronger adsorption. This difference stems from molecular structure and charge distribution. CTAB possesses a single long C16 hydrogen chain and a compact quaternary ammonium headgroup, which allows dense interfacial packing and efficient reduction in surface tension. AOT, on the other hand, has a bulky double-tailed structure with a large sulfosuccinate head carrying two negative charges. Electrostatic repulsion between these doubly charged headgroups and steric hindrance from the twin tails prevent tight packing, producing lower Γ_{max} and larger A_{min} values even in water. The structural difference also modulates their sensitivity to solvent composition: CTAB shows a more pronounced decline in Γ_{max} and a corresponding increase in A_{min} upon adding EG, whereas AOT exhibits a somewhat smaller relative change. This implies that cationic surfactants are more susceptible to polarity and solvation changes introduced by EG, while already strongly hydrated anionic AOT is less affected.

This trend agrees with the work of Ruiz et al. [22], who showed that the addition of EG reduces the surface activity of Triton X-100 by decreasing Γ_{max} and increasing A_{min}. This

https://doi.org/10.3390/colloids10010008

effect is attributed to EG's structure-breaking ability and its interactions with the surfactant, which alter the solvation layer and enhance steric repulsions at the air–liquid interface. These findings further support the observed decrease in surfactant adsorption and looser molecular packing as EG content increases in the solution.

3.2. Contact Angle and Wettability

The spreading of liquid over a solid surface is described as wetting. Situationally, it can also be the penetration of liquid into a porous medium. Quantitatively, the wettability is calculated using the measurement of contact angle (CA), denoted by θ. Geometrically, it is defined as the angle between the liquid phase and the solid phase when these phases are in contact with the gaseous phase, the values of which let us know the extent of wettability, as shown in Scheme 2.

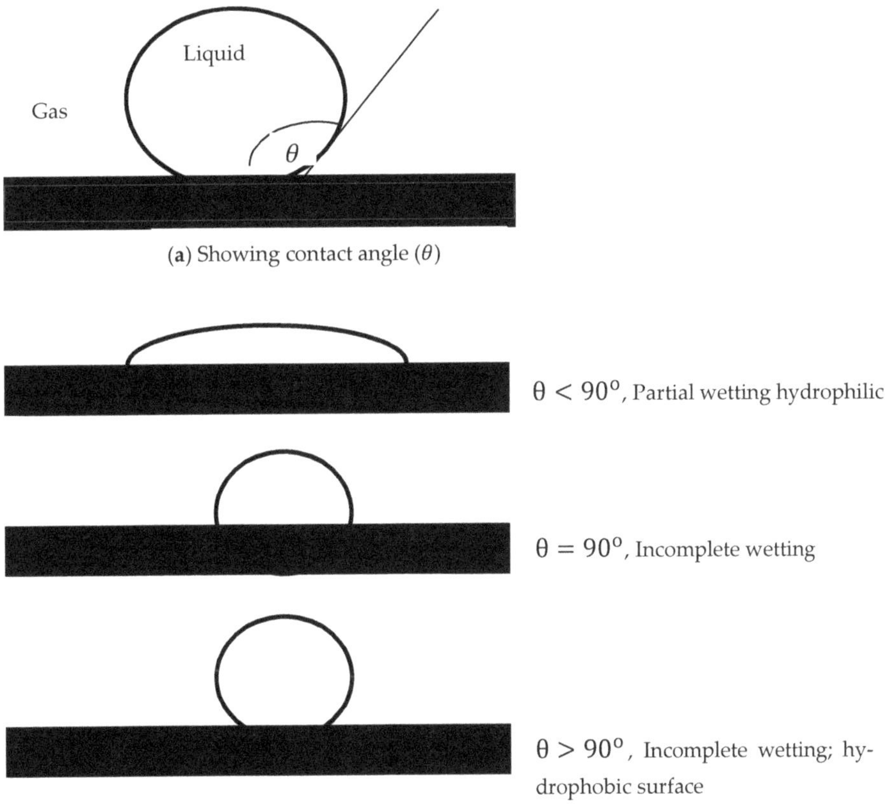

(**b**) Showing wettability at different values of θ

Scheme 2. Schematic representation of wetting regimes based on contact angle.

Surfactants are also used as wetting agents, as they lower the surface tension of a liquid by adsorption at the air–liquid interface; at the same time, they also adsorb at the solid–liquid interface.

Mathematically, CA on the solid surface is related to interfacial tension by Young's Equation (3)

$$cos\theta = \frac{\gamma_{SG} - \gamma_{SL}}{\gamma_{LG}} \quad (3)$$

https://doi.org/10.3390/colloids10010008

where γ_{SG}, γ_{SL}, and γ_{LG} are interfacial tensions between the solid–gas, solid–liquid, and liquid–gas interfaces, respectively.

Surface excess concentration at the gas–liquid interface is calculated using Equation (4)

$$\Gamma_{LG} = -\frac{1}{20303nRT}\left[\frac{d\gamma_{LG}}{dlogC}\right]_{T,P} \tag{4}$$

The Lucassen–Reynolds equation is used to show the relation of surface excess concentrations at three interfaces (Γ_{LG}, Γ_{SL}, and Γ_{SG}) with their respective interfacial tensions as shown below.

$$\frac{\Gamma_{SG} - \Gamma_{SL}}{\Gamma_{LG}} = \frac{d(\gamma_{SG} - \gamma_{SL})}{d\gamma_{LG}} = \frac{d\gamma_{LG}Cos\theta}{d\gamma_{LG}} \tag{5}$$

Assuming $\Gamma_{SG} = 0$, the ratio of Γ_{SL} and Γ_{LG} can be obtained from the slope of a plot of $\gamma_{LG}cos\theta$, known as adhesion tension (A_T), and γ_{LG}.

Work of Adhesion

This is defined by the reversible work required to separate a unit area of liquid from a solid surface. It measures the interactive forces between the two different phases (solid and liquid). It can be obtained from CA using Equation (6) [2]

$$W_A = \gamma_{LG}(1 + cos\theta) \tag{6}$$

For $\theta = 0^o$, $W_A = 2\gamma_{LG}$. This means that the attraction between liquid–solid is equal to or greater than that between liquid–liquid.

Among the additives, alcohol is considered a special additive used in surfactant solutions in physicochemical investigations.

3.2.1. Adhesion Tension vs. Surface Tension—Effect of Ethylene Glycol

Figures 4–7 illustrate how adding ethylene glycol (EG) influences the relationship between A_T and γ for AOT and CTAB surfactant solutions on PTFE and wax surfaces in water and various volume percentages of EG at 298 K. The slope of A_T and γ Equations (4) and (5) plot reflects how the contact angle varies with changing liquid surface tension due to surfactant adsorption at the interfaces. On hydrophobic surfaces, the slope is typically negative [2]. A slope of -1 signifies equal surfactant adsorption at the solid–liquid and air–liquid interfaces, which in turn means the work of adhesion and contact angle remain essentially constant as surface tension changes [24]. On low-energy surfaces like PTFE and wax, however, the measured slopes are negative and deviate from -1, indicating a variable contact angle with surfactant concentration. All our systems show a negative slope, confirming that surfactant adsorption lowers the solid–liquid interfacial tension disproportionately to the drop in liquid surface tension, thereby decreasing the contact angle (enhancing wettability). For example, AOT on PTFE exhibits a slope increasing from -1.25 to 1.08 with EG present, while AOT on wax goes from -1.10 to -0.95. These slopes are more negative than unity in magnitude for AOT on PTFE, implying greater surfactant adsorption at the solid–liquid interface than at the air–water interface, causing pronounced contact angle reduction. In contrast, CTAB on PTFE and wax without EG shows gentler slopes around -0.75, well below unity in magnitude, consistent with less surfactant at the solid–liquid interface relative to the air–liquid interface (single-tail CTAB is less effective at wetting hydrophobic solids).

https://doi.org/10.3390/colloids10010008

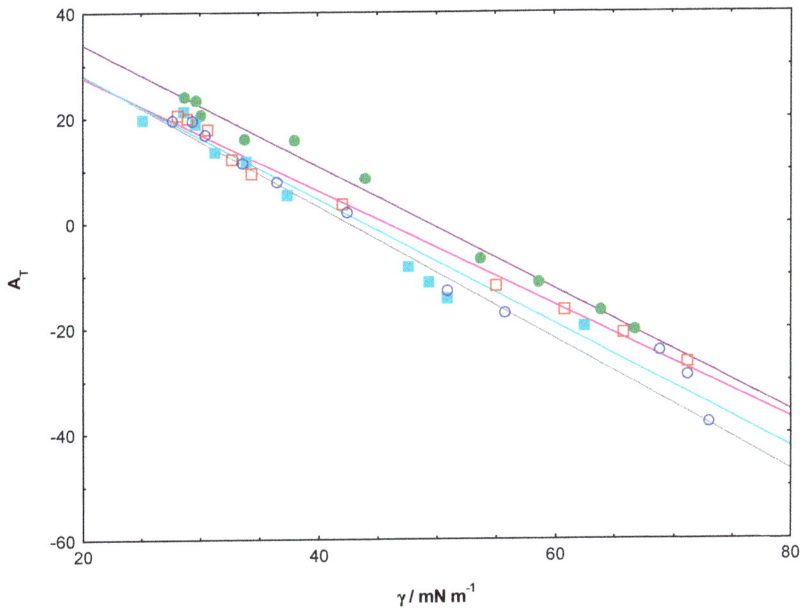

Figure 4. Variation in adhesion tension with Surface tension of AOT on PTFE surface in water (○) and different volume % (5% (□), 10% (•), and 20% (■)) of EG at 298 K.

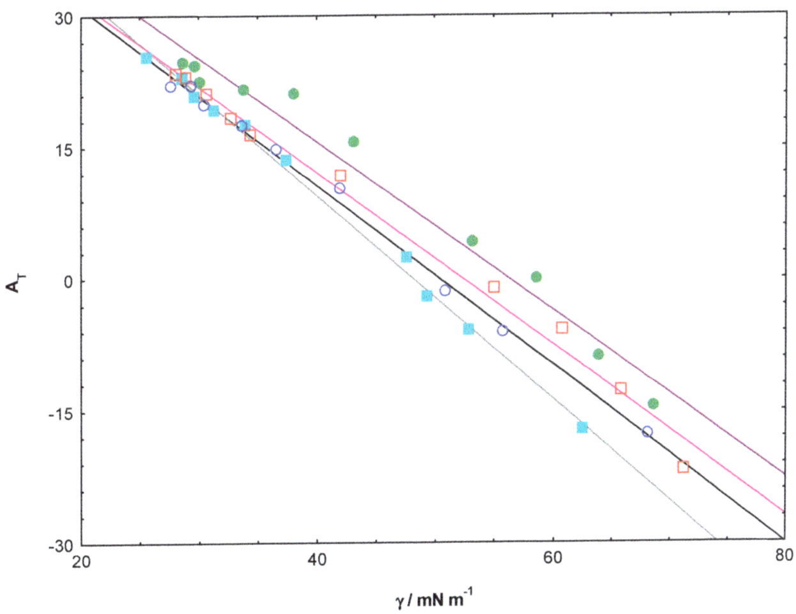

Figure 5. Variation in adhesion tension with Surface tension of AOT on wax surface in water (○) and different volume % (5% (□), 10% (•), and 20% (■)) of EG at 298 K.

https://doi.org/10.3390/colloids10010008

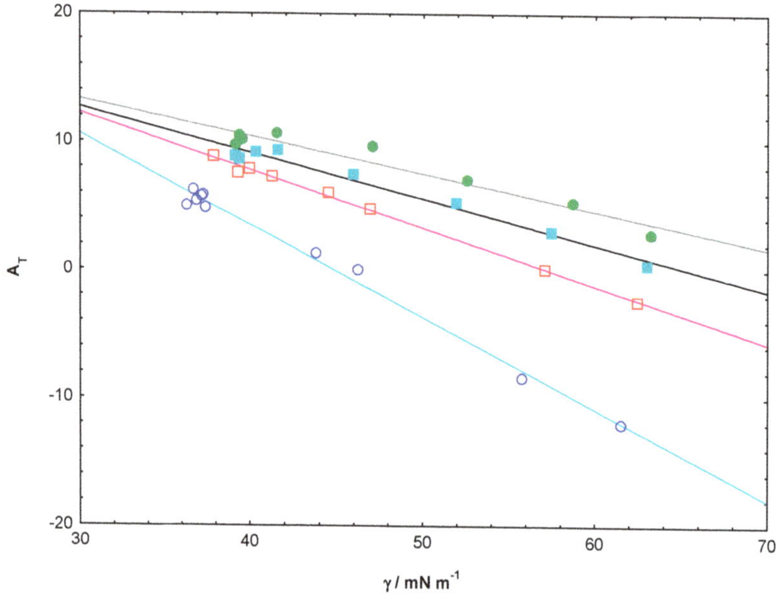

Figure 6. Variation in adhesion tension with Surface tension of CTAB on PTFE surface in water (○) and different volume % (5% (□), 10% (•), and 20% (■)) of EG at 298 K.

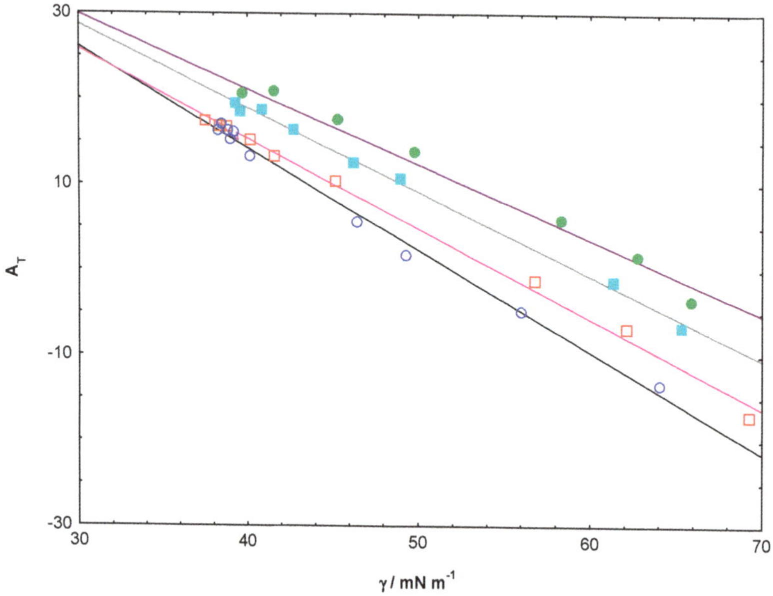

Figure 7. Variation in adhesion tension with Surface tension of CTAB on wax surface in water (○) and different volume % (5% (□), 10% (•), and 20% (■)) of EG at 298 K.

Szymczyk et al. [25] found slopes of −1 for SDs and SDDS on PTFE, indicating equal adsorption densities at the air–liquid interface and PTFE–water interface. AOT's slope of −1.25 to −1.08 aligns with this trend, suggesting strong adsorption at both interfaces. Biswal & Paria [2] reported slopes of −0.80 for Triton X 100 and −0.58 for Igepal CO-630 on PTFE, indicating lower adsorption density at the solid–liquid interface compared

https://doi.org/10.3390/colloids10010008

to the air–liquid interface. CTAB's slope range (−0.75 to −0.29) is consistent with this trend, reflecting weaker adsorption. Chang et al. [26] observed slopes of −0.69 to −0.94 for Gemini surfactants on paraffin (similar to wax), indicating lower adsorption density at the solid–liquid interface compared to the air–liquid interface. AOT's slope range (−1.16 to −0.95) is slightly more negative, suggesting stronger adsorption compared to Gemini surfactants.

3.2.2. Work of Adhesion

Figures 8–11 present W_A as a function of surfactant concentration for both AOT and CTAB on PTFE and wax surfaces in water and various volume % of EG at 298 K. For AOT solutions (Figures 8 and 9), W_A systematically decreases with increasing surfactant concentration across all temperatures and substrates tested. The initial W_A values (at low concentrations) reflect the adhesion of nearly pure water to the hydrophobic surface, which is relatively low due to the energetically unfavorable water-hydrophobe contact. As AOT concentration increases, the work of adhesion decreases, indicating that the surfactant-modified interface requires less energy to maintain contact with the hydrophobic surface.

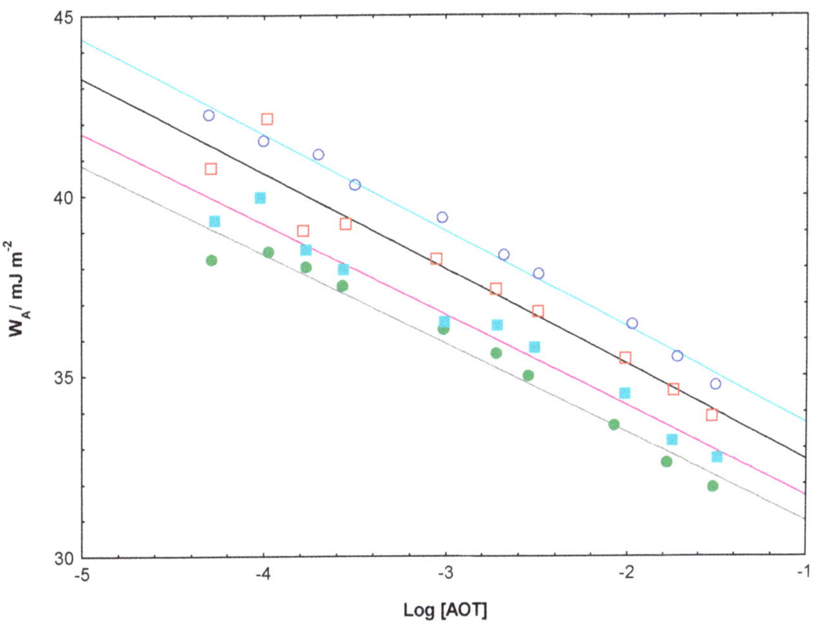

Figure 8. Variation in work of adhesion with logarithmic concentration of AOT on PTFE surface in water (○) and different volume % (5% (□), 10% (•), and 20% (■)) of EG at 298 K.

The decrease in W_A with surfactant concentration can be understood through the lens of interfacial energy balance. Surfactant adsorption at both the liquid-vapor and solid–liquid interfaces reduces the interfacial tensions, thereby reducing the net energy required for adhesion. The temperature dependence shows that higher temperatures generally result in lower W_A values at equivalent concentrations, consistent with enhanced thermal motion and reduced intermolecular cohesion [2].

https://doi.org/10.3390/colloids10010008

the influence of surfactant charge type, headgroup structure, and interfacial packing on wetting outcomes.

Author Contributions: S.K.S. (conceptualization, visualization, writing—original draft, methodology, data curation, supervision); R.B. (visualization, investigation, methodology, writing—original draft); S.G. (methodology, investigation, visualization); P.S. (visualization, methodology, data curation); A.B. (project administration, resources, fund acquisition, writing—reviewing and editing). All authors have read and agreed to the published version of the manuscript.

Funding: This research receives no external funding.

Data Availability Statement: The data supporting the findings of this study are available from the corresponding author upon reasonable request.

Acknowledgments: The authors gratefully acknowledge the Department of Chemistry, Mahendra Morang Adarsh Multiple Campus (Tribhuvan University), Biratnagar, for providing laboratory facilities and research support.

Conflicts of Interest: The authors declare no conflicts of interest.

References

1. Josyula, T.; Malla, L.K.; Thomas, T.M.; Kalichetty, S.S.; Mahapatra, P.S.; Pattamatta, A. Fundamentals and Applications of Surface Wetting. *Langmuir* **2024**, *40*, 8293–8326. [CrossRef]
2. Biswal, N.R.; Paria, S. Wetting of PTFE and Glass Surfaces by Aqueous Solutions of Cationic and Anionic Double-Chain Surfactants. *Ind. Eng. Chem. Res.* **2012**, *51*, 10172–10178. [CrossRef]
3. Shah, S.K.; Das, P.K.; Bhattarai, A. Effect of 2-Propanol on surface properties and wetting behavior of surfactants on the glass surface. *Heliyon* **2025**, *11*, e42352. [CrossRef]
4. Kovalchuk, N.M.; Simmons, M.J.H. Surfactant-mediated wetting and spreading: Recent advances and applications. *Curr. Opin. Colloid Interface Sci.* **2021**, *51*, 101375. [CrossRef]
5. Zdziennicka, A.; Szymczyk, K.; Jańczuk, B. Correlation between surface free energy of quartz and its wettability by aqueous solutions of nonionic, anionic, and cationic surfactants. *J. Colloid Interface Sci.* **2009**, *340*, 243–248. [CrossRef]
6. Menger, F.M.; Rizvi, S.A.A. Relationship between surface tension and surface coverage. *Langmuir* **2011**, *27*, 13975–13977. [CrossRef]
7. Shah, P.; Bhattarai, A. Advances in Surfactants in Foliar Application of Agrochemicals on Mango Leaf Surfaces. *Tenside Surfactants Deterg.* **2020**, *57*, 460–468. [CrossRef]
8. Schenk, J.J.; Becklund, L.E.; Carey, S.J.; Fabre, P.P. What is the "modified" CTAB protocol? Characterizing modifications to the CTAB DNA extraction protocol. *Appl. Plant Sci.* **2023**, *11*, e11517. [CrossRef]
9. Elfeky, S.A.; Mahmoud, S.E.; Youssef, A.F. Applications of CTAB modified magnetic nanoparticles for removal of chromium (VI) from contaminated water. *J. Adv. Res.* **2017**, *8*, 435–443. [CrossRef]
10. Rauniyar, B.S.; Bhattarai, A. Study of conductivity, contact angle, and surface free energy of anionic (SDS, AOT) and cationic (CTAB) surfactants in water and isopropanol mixture. *J. Mol. Liq.* **2021**, *323*, 114604. [CrossRef]
11. Zhang, L.; Wang, Z.L.; Li, Z.Q.; Zhang, L.; Xu, Z.C.; Zhao, S.; Yu, J.Y. Wettability of a quartz surface in the presence of four cationic surfactants. *Langmuir* **2010**, *26*, 18834–18840. [CrossRef]
12. Kwieciński, W.; Segers, T.; van der Werf, S.; van Houselt, A.; Lohse, D.; Zandvliet, H.J.W.; Kooij, S. Evaporation of Dilute Sodium Dodecyl Sulfate Droplets on a Hydrophobic Substrate. *Langmuir* **2019**, *35*, 10453–10460. [CrossRef]
13. Shardt, N.; Bigdeli, M.B.; Elliott, J.A.W.; Tsai, P.A. How Surfactants Affect Droplet Wetting on Hydrophobic Microstructures. *J. Phys. Chem. Lett.* **2019**, *10*, 7510–7515. [CrossRef]
14. Huang, Y.-X.; Wang, Z.; Horseman, T.; Livingston, J.L.; Lin, S. Interpreting contact angles of surfactant solutions on microporous hydrophobic membranes. *J. Membr. Sci. Lett.* **2022**, *2*, 100015. [CrossRef]
15. Jiang, Q.; Du, Y.; Zhang, L.; Ma, W.; Yan, F.; Zhang, L.; Zhao, S. Wettability of a Polymethylmethacrylate Surface by Extended Anionic Surfactants: Effect of Branched Chains. *Molecules* **2021**, *26*, 863. [CrossRef]
16. Bera, B.; Backus, E.H.G.; Carrier, O.; Bonn, M.; Shahidzadeh, N.; Bonn, D. Antisurfactant (Autophobic) Behavior of Superspreader Surfactant Solutions. *Langmuir* **2021**, *37*, 6243–6247. [CrossRef]
17. Ogunmokun, F.A.; Wallach, R. Effect of surfactant surface and interfacial tension reduction on infiltration into hydrophobic porous media. *Geoderma* **2024**, *441*, 116735. [CrossRef]

https://doi.org/10.3390/colloids10010008

18. Zdziennicka, A.; Jańczuk, B. Behavior of cationic surfactants and short-chain alcohols in mixed surface layers at water-air and polymer-water interfaces with regard to polymer wettability. II. Wettability of polymers. *J. Colloid Interface Sci.* **2010**, *350*, 568–576. [CrossRef]
19. Zdziennicka, A.; Jańczuk, B. The adsorption of cetyltrimethylammonium bromide and propanol mixtures with regard to the wettability of polytetrafluoroethylene. II. Adsorption at polytetrafluoroethylene-aqueous solution interface and wettability. *J. Colloid Interface Sci.* **2008**, *318*, 15–22. [CrossRef]
20. Shah, S.K.; Bhattarai, A. Interfacial and Micellization Behavior of Cetyltrimethylammonium Bromide (CTAB) in Water and Methanol-Water Mixture at 298.15 to 323.15 K. *J. Chem.* **2020**, *2020*, 1–13. [CrossRef]
21. Kumar, V.; Patel, P.; Ray, D.; Thareja, P.; Kuperkar, K.; Aswal, V.K.; Bahadur, P. Physiochemical insight into the solution behavior of cationic gemini surfactant in water and ethanol–water systems. *J. Surfactants Deterg.* **2023**, *26*, 623–632. [CrossRef]
22. Ruiz, C.C.; Molina-Bolívar, J.A.; Aguiar, J.; MacIsaac, G.; Moroze, S.; Palepu, R. Thermodynamic and structural studies of Triton X-100 micelles in ethylene glycol-water mixed solvents. *Langmuir* **2001**, *17*, 6831–6840. [CrossRef]
23. Bamyani, N.; Bagheri, A. Interactions between cationic surfactant and polyethylene glycol: Effect of the polymer concentration and alkyl chain length of surfactant. *J. Mol. Liq.* **2025**, *429*, 127598. [CrossRef]
24. Zdziennicka, A.; Krawczyk, J.; Jańczuk, B. Wettability and Adhesion Work Prediction in the Polymer–Aqueous Solution of Surface Active Agent Systems. *Colloids Interfaces* **2018**, *2*, 21. [CrossRef]
25. Szymczyk, K.; Jańczuk, B. Adsorption of Binary Mixtures of Anionic Surfactants at Water–Air and Poly(Tetrafluoroethylene)–Water Interfaces. *J. Surfactants Deterg.* **2010**, *13*, 207–215. [CrossRef]
26. Chang, H.; Cui, Y.; Wang, Y.; Li, G.; Gao, W.; Li, X.; Zhao, X.; Wei, W. Wettability and adsorption of PTFE and paraffin surfaces by aqueous solutions of biquaternary ammonium salt Gemini surfactants with hydroxyl. *Colloids Surf. A Physicochem. Eng. Asp.* **2016**, *506*, 416–424. [CrossRef]

https://doi.org/10.3390/colloids10010008

colloids and interfaces

MDPI

Article

A SERS Substrate for Ultrafast Photosynthetic Au Nanoparticle Growth on WO_3 Nanowires

Shiyong Meng [1,2], Qingsong Deng [1], Lin Zhang [3], Yibo Feng [4], Lei Fan [1], Yuxin Liu [1], Danmin Liu [1,*] and Cong Wang [1,*]

1 Beijing Key Laboratory of Microstructure and Property of Advanced Materials, College of Materials Science and Engineering, Beijing University of Technology, Beijing 100124, China; s202377069@emails.bjut.edu.cn (S.M.); qsdeng@bjut.edu.cn (Q.D.); vanvanl@163.com (L.F.); yuxinliu0923@163.com (Y.L.)
2 College of Physics and Optoelectronic Engineering, Beijing University of Technology, Beijing 100124, China
3 State Key Laboratory of Chemistry for NBC Hazards Protection, Beijing 102205, China; zhanglin_zju@aliyun.com
4 Institute of High Energy Physics, Chinese Academy of Sciences, Beijing 100049, China; fengyb@ihep.ac.cn
* Correspondence: dmliu@bjut.edu.cn (D.L.); smartswang@bjut.edu.cn (C.W.)

Abstract

The practical adoption of surface-enhanced Raman scattering (SERS) technology is often hampered by the high cost, complex fabrication, and poor reproducibility of conventional substrates, which typically rely on noble metals or inefficient semiconductors. Herein, we address key challenges in the practical commercialization of surface-enhanced Raman scattering (SERS) technology by reporting a facile, scalable, and environmentally benign strategy for fabricating a hybrid SERS substrate. This approach integrates Au nanoparticles (NPs) with hydrothermally synthesized WO_3 nanowires through a green photoreduction process, which is rapid, organic-solvent-free, and amenable to large-scale production. The design of the Au/WO_3 nanocomposite capitalizes on the synergistic effect between electromagnetic (EM) enhancement from Au NPs and chemical mechanism (CM) enhancement via charge transfer involving the WO_3 semiconductor. This synergy empowers the substrate with exceptional SERS activity, enabling the sensitive detection of Rhodamine 6G (R6G) down to 10^{-11} M and yielding an enhancement factor (EF) of 4.09×10^6. More importantly, this EM-CM synergy proves critical for detecting molecules with weak affinity, such as the nerve agent simulant dimethyl methylphosphonate (DMMP), achieving a significant signal enhancement of 10^2–10^3 times, which is notably challenging for conventional plasmonic substrates. Beyond sensitivity, the substrate exhibits excellent reproducibility and operational stability, which are paramount for real-world applications. This work presents a nanohybrid strategy that successfully balances scalability, stability, and sensitivity, offering a reliable and cost-effective pathway for advancing SERS technologies toward practical implementation.

Keywords: surface-enhanced Raman scattering (SERS); precious metal–semiconductor hybridization; photosynthesis; ultra-sensitive detection

1. Introduction

Surface-Enhanced Raman Scattering (SERS) represents a groundbreaking spectroscopic methodology, offering exceptional sensitivity and specificity for the detection of trace substances, achieving even single-molecule-level resolution. This technique has found broad utility across biomedical applications (ultra-early tumor marker diagnosis, rapid pathogen identification), environmental monitoring (in situ pollutant detection), materials science (real-time tracking of catalytic interfaces), and public safety (identification of explosives and narcotics). By transcending the sensitivity constraints of conventional Raman spectroscopy, SERS opens new avenues for elucidating molecular behaviors at interfaces [1–12]. The enhancement mechanisms underpinning SERS are predominantly divided into electromagnetic enhancement (EM) and chemical enhancement (CM) [13]. EM stems from localized surface plasmon resonance (LSPR) excited in noble metal nanostructures (Au and Ag nanoparticles) upon laser illumination, generating highly confined electromagnetic "hotspots" at nanogaps or sharp topographic features. These hotspots can amplify Raman scattering by factors of 10^6–10^{10}, with efficiency strongly correlated to the nanostructures' morphology, size, and excitation wavelength [14–16]. Despite their promising performance, conventional noble-metal-based SERS substrates are hampered by intrinsic limitations including high material cost, scarcity, and susceptibility to chemical degradation (oxidation and aggregation), leading to performance fade under prolonged or harsh operational conditions. In light of this, semiconductor-based SERS substrates such as CuO, TiO_2, and ZnO have emerged as promising alternatives [17–20]. These materials operate mainly through the CM mechanism, which involves charge transfer (CT) between analyte molecules and the substrate. Chemical adsorption or strong interfacial interaction facilitates coupling between molecular orbitals and semiconductor band states, modulating molecular polarizability and enhancing the Raman cross-section by approximately 10 to 10^3 times. This enhancement is highly sensitive to the electronic structure alignment at the molecule–substrate interface [21–23], nevertheless, the enhancement factor provided by pure semiconductor substrates remains comparatively modest. Recent research focus has accordingly shifted toward hybrid systems that exploit synergistic effects between EM and CM mechanisms: the plasmonic component provides intense electromagnetic confinement, while the semiconductor facilitates charge transfer and improves chemical specificity [13,24]. This combined approach effectively surmounts the sensitivity limitations of traditional Raman techniques, establishing a robust foundation for single-molecule detection and real-time analysis of interfacial reaction dynamics [25–27].

Against the backdrop of ongoing innovations in semiconductor-based SERS substrates, tungsten trioxide (WO_3) has garnered significant interest owing to its pronounced chemical enhancement properties. Abundant oxygen vacancy defects, a tunable wide bandgap (2.4–2.8 eV), and strong molecular adsorption capacity collectively facilitate efficient charge transfer (CT), establishing WO_3 as a promising platform for SERS applications [28]. Nevertheless, the intrinsic lack of electromagnetic enhancement (EM) in pure WO_3 limits its detection sensitivity. To overcome this limitation, the strategic decoration of gold nanoparticles (Au NPs) onto WO_3 substrates has been proposed, forming Au/WO_3 heterostructures [29]. On one hand, the localized surface plasmon resonance (LSPR) excited in Au NPs generates high-intensity electromagnetic hotspots, offering EM enhancement factors on the order of 10^5–10^7. On the other hand, the Schottky junction formed at the Au–WO_3 interface induces directional electron transfer from WO_3 to Au until their Fermi levels equilibrate. This process creates a built-in electric field at the interface, which efficiently promotes charge separation and optimizes the adsorption energy and polarizability of probe molecules, thereby augmenting the chemical enhancement (CM) by 10–10^3 times [30,31]. Through this synergistic coupling of EM-dominated field amplification and CM-enhanced interfacial interactions, the composite substrate achieves an overall enhancement factor exceeding 10^6, surpassing the performance of either individual component, while combining the stability and cost benefits of WO_3 with the superior EM gain of Au [24,32]; however, traditional Au/WO_3 composite synthesis routes rely on complex polymer stabilizers/capping agents/structure-directing additives [33,34], leading to toxicity [35], SERS deactivation [36], and background noise or unwanted signals in SERS analysis [37]. In this study, we propose an innovative ultraviolet-light-driven in situ reduction strategy that leverages the photocatalytic activity of WO_3 nanowires. Under UV irradiation, photogenerated holes (h^+) in the valence band oxidize sacrificial agents (e.g., methanol), while conduction-band electrons (e^-) directly reduce Au^{3+} ions from $HAuCl_4$ solution, leading to the deposition of Au NPs. This approach enables rapid (within 20 min), organic-free, and in situ deposition of Au NPs onto WO_3 surfaces [38]. By eliminating exogenous reductants and stabilizers, this method ensures an atomically clean Au/WO_3 interface free from molecular contamination. Furthermore, the photo-induced electron flow during synthesis is consistent with the electron transfer direction established in the resultant Schottky junction, which preferentially anchors Au nanoparticles at oxygen vacancy sites and promotes the formation of high-density, well-defined Schottky junctions. This not only enhances charge transfer efficiency but also allows precise regulation of Au NP size and distribution, thereby optimizing LSPR coupling and EM/CM synergy. The resulting SERS substrate exhibits ultrahigh sensitivity supported by dual enhancement mechanisms, offering a novel and scalable route toward industrial fabrication of high-performance detection chips.

Rhodamine 6G (R6G) is widely employed as a standard probe molecule for evaluating SERS substrate performance, owing to its large Raman scattering cross-section, well-defined fingerprint vibrational peaks (612 cm^{-1}, 1360 cm^{-1}, 1510 cm^{-1}), and strong adsorption affinity. Its cationic nature and high photostability ensure robust signal reproducibility, while its suitability for single-molecule detection allows precise assessment of electromagnetic and chemical enhancement (EM/CM) synergy, establishing R6G as an internationally recognized benchmark for quantitative SERS characterization [39]. Dimethyl methylphosphonate (DMMP) serves as an ideal simulant for highly toxic nerve agents such as Sarin and Soman, due to its structural and physicochemical similarity—particularly the presence of phosphoester bonds—coupled with low toxicity. SERS technology meets the critical demands of high sensitivity, specificity, and rapid response required for DMMP detection. As a safe and effective surrogate, DMMP plays an essential role in the development and validation of chemical defense technologies, contributing significantly to national and public security [40].

In this work, WO_3 nanorods were synthesized via a facile hydrothermal route and further functionalized with Au nanoparticles through a UV-induced in situ photoreduction process. This photosynthesis approach is rapid, organic-solvent-free, and readily scalable, providing a green and industrially compatible route for SERS substrate fabrication. The resulting Au/WO_3 substrate exhibits significantly enhanced SERS performance, attributed to the synergistic effect between the electromagnetic enhancement from Au nanoparticles and the chemical enhancement facilitated by the WO_3 semiconductor. The substrate also demonstrates high signal stability, excellent long-term durability, and exceptional reproducibility, as reflected by a relative standard deviation as low as 4.05%. Furthermore, it achieves a detection limit of 10^{-4} M for dimethyl methylphosphonate (DMMP) at room temperature, showcasing its practical potential for the detection of low-adsorption-affinity molecules. This work highlights a feasible and reliable nanohybrid strategy that integrates scalable synthesis, dual enhancement mechanisms, and operational stability, offering a promising pathway toward the practical application of SERS technology.

2. Materials and Methods

All reagents employed in this study are listed in Supplementary Materials Table S1. Unless otherwise specified, all chemicals were used without further purification. The entire experiment was conducted at room temperature, with solutions prepared using deionized water (DI).

2.1. Preparation of Tungsten Oxide

As shown in Scheme S1 (Supporting Information): WCl_6 powder (0.75 g, Aladdin Biochemical Technology Co., Ltd., Shanghai, China) was dissolved in 60 mL of ethanol until the solution turns golden yellow and then transferred to a 100 mL Teflon-lined autoclave. After reacting at 220 °C for 20 h, the deep blue flocculated product was washed repeatedly with ethanol and deionized water at least five times. It was then dried in a 50 °C vacuum oven for 5 h to yield the final powdered product.

2.2. Preparation of Au/WO_3

Disperse 10 mg of the obtained product in a solution of 45 mL water and 5 mL ethanol. Add 0.3 mL chloroauric acid (0.98 wt%, Aladdin Biochemical Technology Co., Ltd., Shanghai, China) and sonicate for 30 min. Expose the mixture to intense UV light irradiation for 900 s. The resulting sample was an orange-red product. After multiple centrifugation washes with ethanol and deionized water, it is dried in a 50 °C vacuum oven for 3 h to obtain the final product.

2.3. SERS Measurement

Silicon Wafer Cleaning: Place several 0.5 × 0.5 cm silicon wafers in ethanol (≥99.7%) and deionized water for repeated cleaning at least five times. Apply ultrasonic treatment during each cleaning cycle to remove residual organic matter from the surface.

Substrate Sol–Gel Coating: Secure the silicon wafers onto transparent microscope slides. Dispense 10 μL of WO_3 and Au/WO_3 sol–gel onto each wafer and allow to dry naturally at room temperature until evaporation is complete.

Adsorption of target molecules: Add 10 μL of R6G (Aladdin Biochemical Technology Co., Ltd., Shanghai, China) ethanol solution at different concentrations and dimethyl methylphosphonate (Aladdin Biochemical Technology Co., Ltd., Shanghai, China) to separate dried samples. Allow natural evaporation at room temperature to ensure tight binding of target molecules to the WO_3 sol (For SERS measurements, prepare a 0.1 M stock solution of the analyte in ethanol. Dilute the stock solution to obtain analyte solutions in the concentration range of 10^{-4} to 10^{-11} M. Store these solutions protected from light at 4 °C until use.).

The R6G/substrate and DMMP/substrate samples were placed under laser micro-area confocal Raman spectroscopy (Renishaw inVia plc, Wotton-under-Edge, Gloucestershire, UK) at 532 nm/785 nm excitation, with light intensities of 1% and 10% respectively, and exposure times of 10 s each for Raman spectroscopy detection. The 520 cm^{-1} peak corresponds to the silicon substrate. To prevent interference from the silicon peak, analysis was confined to the region beyond 600 cm^{-1}.

3. Results and Discussion

3.1. Morphology and Characterization of WO_3 and Au/WO_3

In Figure 1A the phase structures of the as-prepared WO_3 and Au/WO_3 samples were characterized by X-ray diffraction (XRD). The diffraction pattern of WO_3 exhibits major peaks at 2θ = 23.2°, 26.5°, 35.5°, 47.2°, and 55.4°, which correspond to the (002), (120), (121), (004), and (420) crystal planes of monoclinic WO_3 (JCPDS 00-020-1324), respectively. For the Au/WO_3 composite, additional distinct peaks are observed at 2θ = 38.2°, 44.3°, 64.7°, and 77.7°, which can be assigned to the (111), (200), (220), and (311) planes of face-centered cubic Au (JCPDS 03-065-8601) [41]. These XRD results confirm the successful formation of the Au/WO_3 heterostructure without altering the crystalline framework of WO_3. The morphology and elemental composition of the as-synthesized Au/WO_3 composite were characterized using scanning electron microscopy (SEM) (Qunata600F FEI Company, Hillsboro, OR, USA), transmission electron microscopy (TEM) (FEI Talos F200X Thermo Fisher Scientific, Hillsboro, OR, USA), and energy-dispersive X-ray spectroscopy (EDX) (FEI Talos F200X Thermo Fisher Scientific, Hillsboro, OR, USA). As shown in Figure S1, SEM images confirm the successful formation of WO_3 nanorods, along with Au/WO_3 composites with varying Au loadings. It is evident that the density of deposited Au nanoparticles exhibits a clear positive correlation with the concentration of chloroauric acid used during synthesis. Figure 1B,C present TEM and HAADF-STEM images that confirm the successful synthesis of WO_3 nanowires and the subsequent uniform loading of Au nanoparticles onto the WO_3 nanorods. The average particle size of the Au NPs is approximately 13 nm, as further quantified by the size distribution histogram in Figure S2. The increased Au loading capacity is demonstrated in Figure S3, where a corresponding darkening of the substrate color visually reflects the higher nanoparticle coverage. For more precise structural characterization, high-resolution transmission electron microscopy (HR-TEM) was employed to examine the atomic-scale morphology of the Au/WO_3 heterostructure. As shown in Figure 1D, the measured lattice spacings of 0.376 nm and 0.232 nm are assigned to the (200) plane of the WO_3 and the (200) plane of Au NPs [42,43] respectively. Furthermore, HAADF-STEM imaging (Figure 1E) provides clearer visualization of the distributed Au NPs, while elemental mapping of W, O, and Au conclusively confirms the successful deposition of Au NPs onto the WO_3 support.

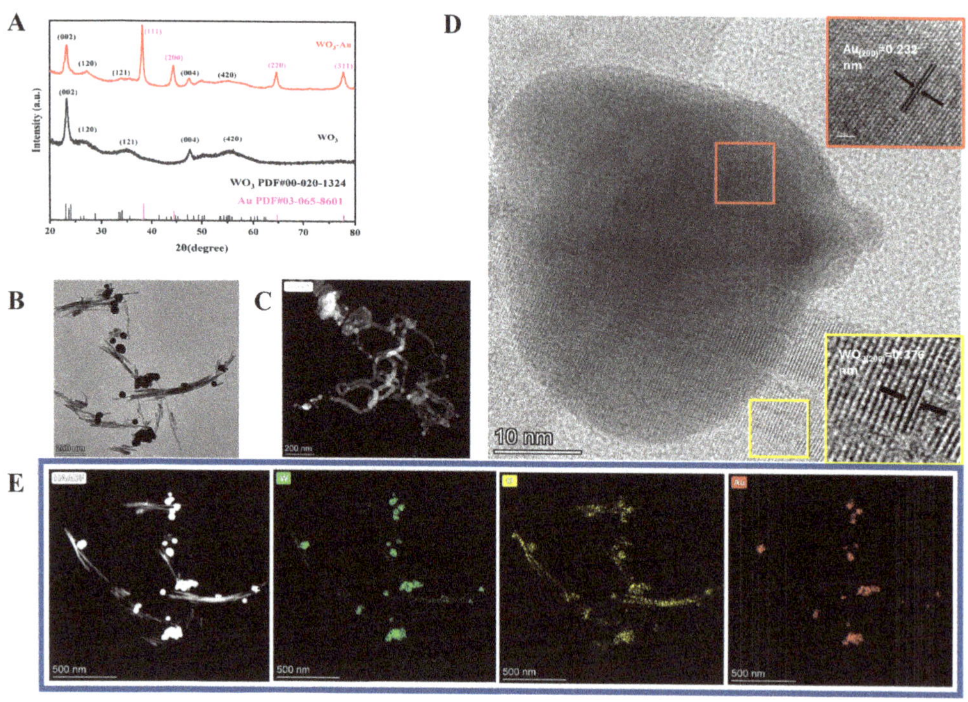

Figure 1. (**A**) XRD patterns of Au/WO_3 and WO_3; (**B**) TEM image of Au/WO_3; (**C**) HAADF-STEM image of Au/WO_3; (**D**) HR-TEM image of Au/WO_3; (**E**) HAADF-STEM image of Au/WO_3; STEM-EDS mapping images of W, O, and Au.

3.2. SERS Performance of WO_3 and Au/WO_3 SERS Substrates

The surface plasmon resonance (SPR) peak of Au nanoparticles lies close to the excitation wavelength of 532 nm, favoring the generation of strong localized electric fields and thereby enhancing Raman signals [44]. Accordingly, an excitation wavelength of 532 nm was selected for the detection of R6G in this study. Rhodamine 6G was employed as a probe molecule to evaluate the SERS performance of both WO_3 and Au/WO_3 substrates, elucidating the effect of Au nanoparticle deposition on the Raman detection sensitivity. As illustrated in Figure S4, varying the amount of tetrachloroauric acid solution during synthesis—from low to high concentrations—resulted in an initial marked enhancement in the Raman peak intensity of R6G, followed by a decline as the Au loading continued to increase. This trend suggests that the Au nanoparticle coverage is initially insufficient for optimal SERS enhancement, but becomes excessive at higher loadings, likely due to aggregation or screening effects. Based on these observations, an intermediate Au loading amount was selected for all subsequent experiments. Figure 2A,B present the Raman spectra of R6G acquired on WO_3 and Au/WO_3 substrates, respectively, across a range of concentrations (from 10^{-4} to 10^{-11} mol/L) under consistent experimental conditions. (The assignment of characteristic R6G Raman peaks and their corresponding vibrational modes are summarized in Table S2 [45]). For the WO_3 substrate, the signal at 612 cm^{-1} became indistinguishable from the background at an R6G concentration of 10^{-7} mol/L, indicating a detection limit of approximately 10^{-7} M. In contrast, distinct Raman signals were still detectable on the Au/WO_3 substrate even at an R6G concentration as low as 10^{-11} mol/L, demonstrating a detection limit below 10^{-11} M. The deposition of Au nanoparticles thus enhances the detection sensitivity by four orders of magnitude. The enhancement factors (EF) for both substrates were quantified using a pristine SiO_2 substrate as reference. As detailed in the Supporting Information, the EF values were calculated to be 1.5×10^5 for WO_3 and 4.09×10^6 for Au/WO_3—indicating an order-of-magnitude improvement in SERS performance with Au modification. This enhancement exceeds that of most semiconductor-based SERS substrates reported previously, as compiled in Table S3. The uniformity and long-term stability of SERS substrates remain critical challenges for their practical application. To assess spatial uniformity, SERS mapping was performed at 20 randomly selected points on the Au/WO_3 substrate. The relative standard deviation (RSD) of the peak intensity at 612 cm^{-1} was calculated to be 4.05% and 6.89% for different R6G concentrations in Figures 2C and S5, indicating excellent signal homogeneity. Furthermore, the substrate exhibited remarkable long-term stability over a 100-day period (Figure 2D). The SERS intensity retained 86% of its original value after 50 days and 75% after 100 days, demonstrating minimal degradation. Quantitative analysis was conducted based on the characteristic R6G peak at 612 cm^{-1}, which is attributed to the in-plane bending vibration of the C-C-C ring. This peak remained clearly detectable even at a concentration as low as 10^{-11} M. As shown in Figure 2E, the SERS intensity decreased progressively with R6G concentration from 10^{-4} to 10^{-11} M. A linear relationship was established with the regression equation logI(612 cm^{-1}) = 5.679 logR6G + 0.30208, yielding a high correlation coefficient (R^2) of 0.97616, which confirms the excellent quantitative detection capability of the Au/WO_3 substrate.

Subsequently, we evaluated the SERS performance of the substrate for detecting dimethyl methylphosphonate (DMMP). In real-world environments such as those containing airborne particulates, fabric surfaces, skin secretions, or plastic residues, strong fluorescent background interference is often encountered. To mitigate this, a laser excitation wavelength of 785 nm was selected, as it effectively suppresses fluorescence and enables the acquisition of well-resolved characteristic fingerprint Raman spectra of DMMP across diverse conditions. As shown in Figure 2F, Raman spectra of DMMP at various concentrations (from 1 to 10^{-4} mol/L) were collected on both WO_3 and Au/WO_3 substrates under consistent experimental parameters. (Assignments of characteristic DMMP Raman peaks and corresponding vibrational modes are provided in Table S3 [46]). The characteristic peak at 710 cm^{-1} became undetectable on pure WO_3 at a DMMP concentration of 10^{-2} mol/L, indicating a detection limit of approximately 10^{-2} M. In contrast, distinct Raman signals were still observable on the Au/WO_3 substrate even at a DMMP concentration as low as 10^{-4} mol/L, demonstrating a detection limit below 10^{-4} M. The incorporation of Au nanoparticles thus enhanced the detection sensitivity for DMMP by 2–3 orders of magnitude. Furthermore, the WO_3 and Au/WO_3 substrate itself exhibits a clean spectral background in the region of interest, and its SERS signal for DMMP remains stable against common environmental interferents, including variations in relative humidity and the presence of solvent vapors, as systematically validated in Figure S6. These results confirm that the presence of Au NPs significantly facilitates molecular detection, enabling higher quantitative accuracy and excellent reproducibility.

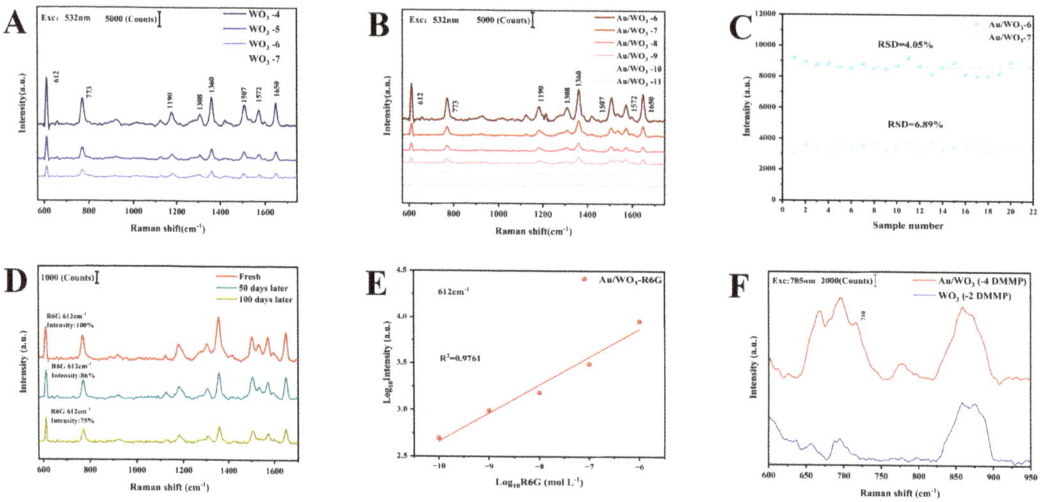

Figure 2. (**A**,**B**) show the R6G test SERS spectra on Au/WO_3 and WO_3 substrates; (**C**) Uniformity of R6G Raman signals at 20 random positions on Au/WO_3; (**D**) Stability test of Au/WO_3; (**E**) Calibration curves of the logarithmic Raman peak (612 cm^{-1}) versus the logarithmic concentration of R6G on Au/WO_3 SERS substrate; (**F**) DMMP test SERS spectra of Au/WO_3 and WO_3 substrates.

3.3. *Mechanism Exploration*

Surface chemical composition and electronic structure of the substrates were investigated by X-ray photoelectron spectroscopy (XPS). All XPS spectra were charge-corrected by referencing the C 1s peak to 284.8 eV, with the corresponding C 1s spectrum provided in Figure S7. Figure 3A presents the survey spectra of both WO_3 and Au/WO_3, confirming the successful incorporation of Au through the emergence of Au 4f peaks in the composite material. High-resolution spectra of the W 4f region are displayed in Figure 3B,C. For pristine WO_3, the spin-orbit doublet corresponding to W $4f_{7/2}$ and W $4f_{5/2}$ appeared at binding energies of 35.26 eV and 37.39 eV, respectively. After Au deposition, these peaks shifted to 35.49 eV and 37.65 eV, representing a positive shift of approximately 0.25 eV. This increase in binding energy suggests electron transfer from W to Au, resulting in reduced electron density around tungsten nuclei and decreased shielding of core electrons, consistent with the formation of a Schottky junction at the Au/WO_3 interface [47]. As further supported by the deconvolution of the high-resolution W 4f spectrum (Figure S7), the appearance of W^{5+} species following Au deposition indicates an increased presence of oxygen vacancies [48]. Figure 3D,E show the corresponding O 1s spectra, which exhibit noticeable asymmetry and broadening. Each spectrum was fitted with two components corresponding to lattice oxygen (W–O) at lower binding energy and surface-adsorbed oxygen (e.g., –OH or adsorbed H_2O) at higher binding energy. The binding energies of the O 1s peaks in WO_3 were observed at 530.11 eV and 530.97 eV, while those in Au/WO_3 shifted to 530.36 eV and 531.50 eV. Moreover, the relative area of the high-binding-energy component (surface-adsorbed oxygen) increased significantly after Au modification, indicating a rise in surface oxygen vacancy defects. This finding aligns well with the W 4f XPS result. Numerous studies have established that oxygen vacancies can effectively enhance the SERS performance of semiconductor substrates [49–52]. XPS analysis in this work indicates that, compared to pure WO_3, the higher concentration of oxygen defects in Au/WO_3 introduces dangling bonds and induces band tailing, driving the system into a metastable state. This electronic configuration facilitates the escape and transfer of surface electrons. The resulting electron delocalization promotes more efficient interfacial charge transfer, thereby significantly improving SERS activity [53–55]. As shown in Figure 3F, the Au 4f spectrum displays two well-defined peaks at binding energies of 83.71 eV and 87.32 eV, corresponding to Au^0 $4f_{7/2}$ and $4f_{5/2}$, respectively, confirming the metallic nature of the deposited Au nanoparticles on the WO_3 nanostructure.

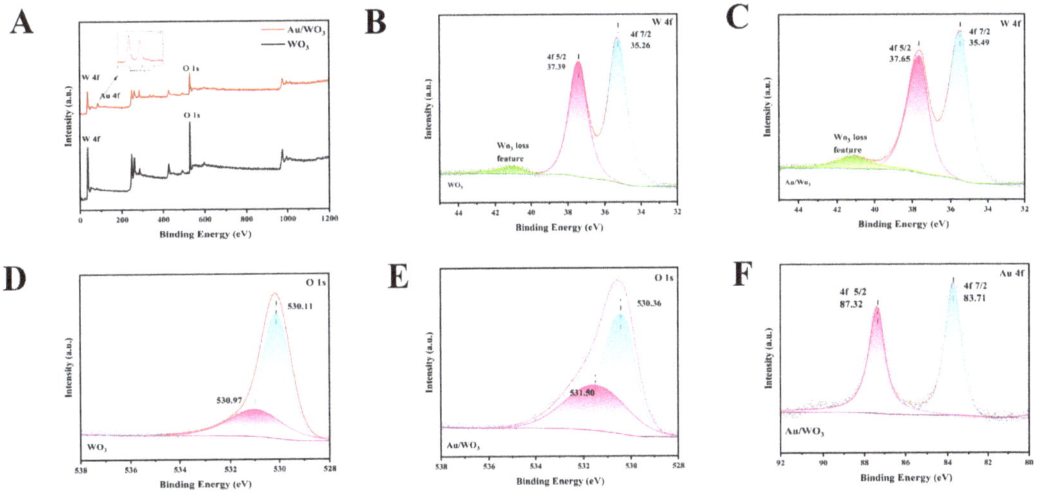

Figure 3. (**A**) Full XPS spectra of Au/WO_3 and WO_3 substrates; (**B**,**C**) High-resolution XPS W4f spectra of WO_3 and Au/WO_3; (**D**,**E**) High-resolution XPS O1s spectra of WO_3 and Au/WO_3; (**F**) High-resolution XPS Au4f spectrum of Au/WO_3.

To elucidate the interfacial binding chemistry, particularly in the absence of conventional capping agents, the surface charge of the Au/WO_3 composite was investigated. Zeta potential measurement confirms that the substrate possesses a negative surface charge (see Table S5). This inherent negativity is crucial for the initial electrostatic adsorption of cationic probe molecules like R6G. We attribute this negative charge to the adsorption of anionic species, primarily chloride ions (Cl^-) and possibly gold-chloro complexes, which are inherent by-products from the hydrolysis and photoreduction of the $HAuCl_4$ precursor [56,57]. These species adsorb onto the Au NP surface during the in situ growth, forming a negative electrical layer that functionally mimics the role of citrate in traditional syntheses, thereby facilitating the electrostatic capture of positively charged molecules and contributing to nanoparticle stabilization.

As shown in Figure 4A,B, the experimental results reveal that Au/WO_3 exhibits enhanced optical absorption and a reduced bandgap compared to pure WO_3, accompanied by a distinct localized surface plasmon resonance (LSPR) absorption feature [58]. To ensure the accuracy and reproducibility of the bandgap determination, the Tauc plots were quantitatively analyzed. The optical bandgap (E_g) was determined by applying a linear fit to the rising edge of the $(\alpha h\nu)^2$ versus photon energy (hν) plot using a nonlinear fitting tool in Origin Lab software. The fitting interval was carefully selected based on the linearity of the data points, ensuring a high coefficient of determination ($R^2 > 0.99$). The bandgap value was then obtained by extrapolating the fitted line to the intercept with the hν axis, where $(\alpha h\nu)^2 = 0$. This quantitative fitting method yielded bandgaps of 2.94 eV for pristine WO_3 and 2.62 eV for Au/WO_3. We note that these values are in close agreement (within 0.01 eV) with our initial estimates, confirming the robustness of the reported bandgap trend and values, while providing a more objective and rigorous foundation. It is well established that a narrower bandgap promotes more efficient visible-light absorption, leading to increased generation of photogenerated charge carriers. This effect facilitates improved charge transfer processes, thereby significantly enhancing the Raman signal intensity. Simultaneously, the capture of photoexcited carriers within the substrate effectively reduces noise arising from resonant Raman scattering [25]. Recent studies have revealed that a portion of the observed SERS signal may originate from photoluminescence (PL) modulated by noble metal nanostructures. However, due to photon scattering mechanisms, PL typically contributes to an unwanted background interference in SERS measurements [59]. Interestingly, Yang et al. elucidated the role of charge transfer in Nd-doped ZnO systems, demonstrating a direct correlation between PL quenching and SERS enhancement [60]. As shown in the PL spectrum in Figure 4C, a significant fluorescence quenching effect is observed in Au/WO_3 compared to pure WO_3. The quenching peak aligns with the plasmon resonance band of Au NPs, confirming the contribution of an electromagnetic enhancement mechanism. The suppression of PL emission helps mitigate interference from substrate-originated luminescence, which is commonly perceived as Raman noise. The reduced background facilitates the detection of weak molecular Raman signals, which is particularly crucial for trace analyte detection. Therefore, surface modification with Au NPs enhances the detection sensitivity not only by reducing the recombination rate of photogenerated charge carriers but also by minimizing spectral background interference.

The electronic band structures of WO_3 and Au/WO_3 were further probed using valence band X-ray photoelectron spectroscopy (VB-XPS), as depicted in Figure 4D. The valence band maximum (VBM) was measured at 1.31 eV for pristine WO_3 and shifted to 1.52 eV after Au deposition, indicating a modification of the electronic environment. To thermodynamically rationalize the charge transfer process, we evaluated potential charge transfer pathways between the substrate and Rhodamine 6G (R6G) molecules, with a schematic illustration provided in Figure 4E. Based on Tauc plot analysis and VB-XPS results, the conduction band (CB) and valence band (VB) positions of WO_3 were determined to be −3.53 eV and −6.47 eV, respectively. The Fermi level (E_F) of Au is −5.1 eV [61,62], while the highest occupied molecular orbital (HOMO) and the lowest unoccupied molecular orbital (LUMO) of R6G lie at −5.7 eV and −3.4 eV, respectively. Under 532 nm laser excitation, four distinct resonance mechanisms contribute synergistically to the chemical enhancement: molecular resonance (μmol) within R6G, exciton resonance (μex) in WO_3, photoinduced charge transfer (μPICT) between the semiconductor and molecular energy levels, and plasmon-induced hot electron transfer (μPHET). The energy differences from the Fermi level of Au and the VB of WO_3 to the LUMO of R6G are 1.7 eV and 3.07 eV, respectively, enabling direct electron transfer from both Au and WO_3 to the LUMO of R6G under laser irradiation. This multi-pathway charge injection considerably increases the electron transfer probability and enhances the SERS activity. Additionally, the energy offset between the Fermi level of Au and the CB of WO_3 is 1.57 eV, permitting efficient electron transfer from Au to the CB of WO_3, thereby further improving charge separation and transfer efficiency. The collective contribution of these interfacial charge transfer processes significantly enhances the SERS performance. Furthermore, the localized electric field generated by Au nanoparticles amplifies the polarization effect resulting from charge transfer, leading to additional SERS enhancement. Through the combined electromagnetic and chemical enhancement mechanisms, the charge transfer between R6G and the Au/WO_3 composite is markedly facilitated, resulting in a high enhancement factor of 4.09×10^6 and a detection limit as low as 10^{-11} M.

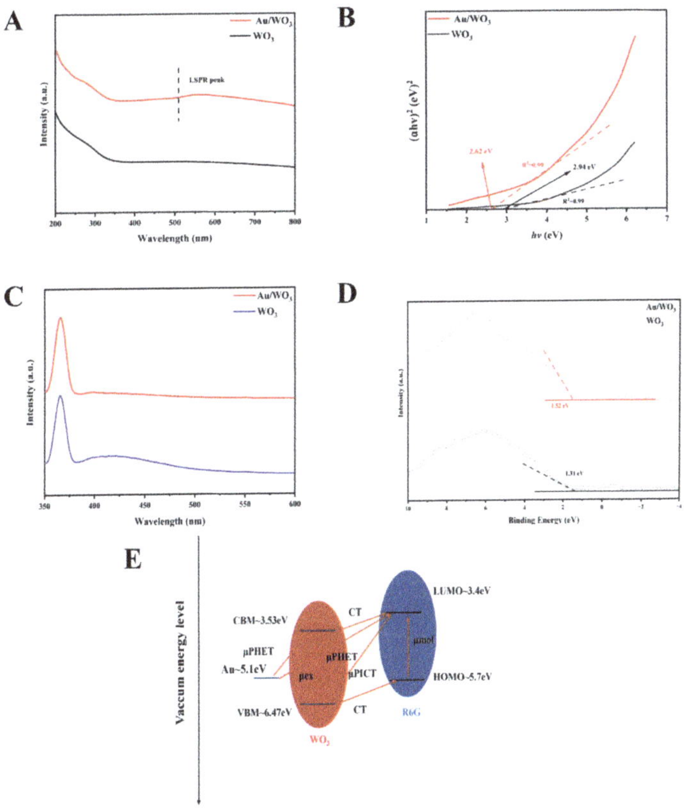

Figure 4. Band structure analysis. (**A**) UV-Vis absorption spectrum. (**B**) Tauc plot. (**C**) Steady-state photoluminescence of Au/WO_3 and WO_3. (**D**) XPS-VB mapping. (**E**) Schematic illustrating the possible charge transfer process between Au/WO_3 and R6G upon 532 nm excitation.

4. Conclusions

This work presents a rationally designed Au/WO_3 hybrid substrate for surface-enhanced Raman scattering (SERS), fabricated through a green and scalable UV-induced photoreduction strategy. This approach enables the additive-free, in situ deposition of Au nanoparticles onto WO_3 nanowires, ensuring an atomically clean interface and offering a viable route toward large-scale production. The exceptional SERS performance of the substrate stems from the synergistic interplay between electromagnetic enhancement from

Au nanoparticles and chemical enhancement facilitated by the WO_3 support, further amplified by the formation of a Schottky junction and an increased concentration of oxygen vacancies. The substrate demonstrates remarkable operational stability, maintaining 75% of its initial SERS activity after 100 days, along with excellent reproducibility (relative standard deviation of 4.05%). It achieves highly sensitive detection of Rhodamine 6G down to 10^{-11} M and reliably detects the nerve agent simulant DMMP at 10^{-4} M, confirming its practical potential for trace analyte sensing. This study not only provides fundamental insights into the charge transfer mechanisms in semiconductor–metal heterostructures but also establishes a robust and eco-friendly fabrication pathway, highlighting a significant step toward the real-world application of SERS technology in environmental and security monitoring.

Supplementary Materials: The following supporting information can be downloaded at: https://www.mdpi.com/article/10.3390/colloids9050070/s1, Figure S1. SEM images of (A) WO_3, (B) 150 μL $HAuCl_4$ Au/WO_3 and (C) 300 μL $HAuCl_4$ Au/WO_3 (D) 450 μL $HAuCl_4$ Au/WO_3; Figure S2. Size distribution histograms of Au NPs on WO_3; Figure S3. Photos taken by iphone of (A) 150 μL $HAuCl_4$ Au/WO_3, (B) 300 μL $HAuCl_4$ Au/WO_3 and (C) 450 μL $HAuCl_4$ Au/WO_3; Figure S4. SERS performance diagrams with different contents of Au nanoparticles; Figure S5. The uniformity of R6G Raman signals of Au/WO_3 at 20 random positions; Figure S6. (A) Raman peak background of WO_3 and Au/WO_3 at 785 nm, (B) SERS Performance of DMMP under Different Solvent Conditions, (C) SERS Performance of DMMP under Different Solvent Conditions; Figure S7. The XPS W4f spectrum of (A) WO_3; (B) Au/WO_3, The XPS C1s spectrum of (C) WO_3, (D) Au/WO_3; Table S1. Experimental reagents; Table S2. Mode assignment of the Raman peaks for R6G; Table S3. Comparison of EFs and limits of detection (LODs) of various semiconductors with SERS activity in the literature (R6G); Table S4. Mode assignment of the Raman peaks for DMMP; Table S5. Zeta potential; Scheme S1. Synthesis diagram of Au/WO_3 species [63–68].

Author Contributions: Conceptualization, S.M.; methodology, S.M.; validation, C.W. and D.L.; investigation, S.M. and Y.F.; writing—original draft preparation, Q.D. and Y.L.; writing—review and editing, Q.D., L.Z. and L.F.; supervision, C.W.; funding acquisition, C.W. All authors have read and agreed to the published version of the manuscript.

Funding: This work was supported by the National Natural Science Foundation of China (Grant Nos. 12574003, 52471250), Beijing Natural Science Foundation (Grant Nos. L248027 and L245019).

Data Availability Statement: The original contributions presented in this study are included in the article. Further inquiries can be directed to the corresponding authors.

Conflicts of Interest: The authors declare no conflict of interest.

References

1. Hassanain, W.A.; Izake, E.L. Toward Label-Free SERS Detection of Proteins through Their Disulfide Bond Structure. *SLAS Discov. Adv. Sci. Drug Discov.* **2020**, *25*, 87–94. [CrossRef] [PubMed]
2. Feng, R.; Miao, Q.; Zhang, X.; Cui, P.; Wang, C.; Feng, Y.; Gan, L.; Fu, J.; Wang, S.; Dai, Z.; et al. Single-atom sites on perovskite chips for record-high sensitivity and quantification in SERS. *Sci. China Mater.* **2022**, *65*, 1601–1614. [CrossRef]
3. Feng, R.; Fu, S.; Liu, H.; Wang, Y.; Liu, S.; Wang, K.; Chen, B.; Zhang, X.; Hu, L.; Chen, Q.; et al. Single-Atom Site SERS Chip for Rapid, Ultrasensitive, and Reproducible Direct-Monitoring of RNA Binding. *Adv. Healthc. Mater.* **2024**, *13*, e2301146. [CrossRef] [PubMed]
4. Wang, K.; Gao, Y.; Fang, Z.; Li, H.; Feng, R.; Wang, Y.; Feng, Y.; Li, W.; Zhang, S.; Hu, L.; et al. SERS detection for pesticide residue via a single-atom sites decoration strategy. *Appl. Surf. Sci.* **2023**, *621*, 156832. [CrossRef]
5. Dai, X.; Fu, W.; Chi, H.; Mesias, V.S.D.; Zhu, H.; Leung, C.W.; Liu, W.; Huang, J. Optical tweezers-controlled hotspot for sensitive and reproducible surface-enhanced Raman spectroscopy characterization of native protein structures. *Nat. Commun.* **2021**, *12*, 1292. [CrossRef] [PubMed]

6. Yang, L.; Kim, T.; Cho, H.; Luo, J.; Lee, J.; Chueng, S.D.; Hou, Y.; Yin, P.T.; Han, J.; Kim, J.H.; et al. Hybrid Graphene-Gold Nanoparticle-Based Nucleic Acid Conjugates for Cancer-Specific Multimodal Imaging and Combined Therapeutics. *Adv. Funct. Mater.* **2020**, *31*, 2006918. [CrossRef]
7. Wang, J.; Zhao, J.; Xiao, L.; Zhao, X.; He, S.; Zhou, Y.; Liu, J.; Li, Y.; Peng, L.; Liu, W. Bio-inspired 3D Sub-Microstructures Coupling of Au Nanoparticles for SERS Detection of Trace Fentanyl. *Adv. Funct. Mater.* **2024**, *34*, 2411048. [CrossRef]
8. Wu, L.; Tang, X.; Wu, T.; Zeng, W.; Zhu, X.; Hu, B.; Zhang, S. A review on current progress of Raman-based techniques in food safety: From normal Raman spectroscopy to SESORS. *Food Res. Int.* **2023**, *169*, 112944. [CrossRef]
9. Afroozeh, A. A Review of Developed Surface-Enhanced Raman Spectroscopy (SERS)-Based Sensors for the Detection of Common Hazardous Substances in the Agricultural Industry. *Plasmonics* **2024**, *20*, 63–81. [CrossRef]
10. Li, X.; Xu, C.; Yan, L.; Feng, Y.; Li, H.; Ye, C.; Zhang, M.; Jiang, C.; Li, J.; Wu, Y. A plasmonic AgNP decorated heterostructure substrate for synergetic surface-enhanced Raman scattering identification and quantification of pesticide residues in real samples. *Anal. Methods* **2022**, *14*, 3250–3259. [CrossRef]
11. Ustun, O.; Yilmaz, A.; Yilmaz, M. Catalytic and SERS activities of WO_3-based nanowires: The effect of oxygen vacancies, silver nanoparticle doping, and the type of organic dye. *Phys. Chem. Chem. Phys.* **2022**, *24*, 18615–18626. [CrossRef] [PubMed]
12. Chen, B.; Meng, S.; Liu, D.; Deng, Q.; Wang, C. In Situ SERS Monitoring of Schiff Base Reactions via Nanoparticles on a Mirror Platform. *Catalysts* **2024**, *14*, 803. [CrossRef]
13. Han, X.X.; Rodriguez, R.S.; Haynes, C.L.; Ozaki, Y.; Zhao, B. Surface-enhanced Raman spectroscopy. *Nat. Rev. Methods Prim.* **2021**, *1*, 87. [CrossRef]
14. Zhan, C.; Chen, X.-J.; Yi, J.; Li, J.-F.; Wu, D.-Y.; Tian, Z.-Q. From plasmon-enhanced molecular spectroscopy to plasmon-mediated chemical reactions. *Nat. Rev. Chem.* **2018**, *2*, 216–230. [CrossRef]
15. Ding, S.-Y.; You, E.-M.; Tian, Z.-Q.; Moskovits, M. Electromagnetic theories of surface-enhanced Raman spectroscopy. *Chem. Soc. Rev.* **2017**, *46*, 4042–4076. [CrossRef] [PubMed]
16. Alessandri, I.; Lombardi, J.R. Enhanced Raman Scattering with Dielectrics. *Chem. Rev.* **2016**, *116*, 14921–14981. [CrossRef] [PubMed]
17. Yang, B.; Shao, X.; Gu, X.; Wang, K.; Ning, X.; Xia, J.; Xie, M.; Tang, Y.; Li, Q.; Tian, S. CuO@AgNPs nanozyme cavity arrays on screen-printed electrodes for ultrasensitive and on-site SERS detection. *Chem. Eng. J.* **2023**, *471*, 144522. [CrossRef]
18. Wang, X.; Shi, W.; Jin, Z.; Huang, W.; Lin, J.; Ma, G.; Li, S.; Guo, L. Remarkable SERS Activity Observed from Amorphous ZnO Nanocages. *Angew. Chem. Int. Ed. Engl.* **2017**, *56*, 9851–9855. [CrossRef]
19. Estevez, M.; Alvarez, M.; Lechuga, L. Integrated optical devices for lab-on-a-chip biosensing applications. *Laser Photon-Rev.* **2011**, *6*, 463–487. [CrossRef]
20. Simas, M.V.; Davis, G.A.; Hati, S.; Pu, J.; Goodpaster, J.V.; Sardar, R. Anisotropically Shaped Plasmonic WO_{3-x} Nanostructure-Driven Ultrasensitive SERS Detection and Machine Learning-Based Differentiation of Nitro-Explosives. *ACS Appl. Mater. Interfaces* **2025**, *17*, 11309–11324. [CrossRef]
21. Lombardi, J.R.; Birke, R.L. Theory of Surface-Enhanced Raman Scattering in Semiconductors. *J. Phys. Chem. C* **2014**, *118*, 11120–11130. [CrossRef]
22. Park, W.-H.; Kim, Z.H. Charge Transfer Enhancement in the SERS of a Single Molecule. *Nano Lett.* **2010**, *10*, 4040–4048. [CrossRef]
23. Wang, X.; Guo, L. SERS Activity of Semiconductors: Crystalline and Amorphous Nanomaterials. *Angew. Chem. Int. Ed. Engl.* **2019**, *59*, 4231–4239. [CrossRef]
24. Li, Q.; Wang, J.; Ding, Q.; Chen, M.; Ma, F. Coupling effect on charge-transfer mechanism of surface-enhanced resonance Raman scattering. *J. Raman Spectrosc.* **2017**, *48*, 560–569. [CrossRef]
25. Langer, J.; Jimenez de Aberasturi, D.; Aizpurua, J.; Alvarez-Puebla, R.A.; Auguié, B.; Baumberg, J.J.; Bazan, G.C.; Bell, S.E.J.; Boisen, A.; Brolo, A.G.; et al. Present and Future of Surface-Enhanced Raman Scattering. *ACS Nano* **2020**, *14*, 28–117. [CrossRef]
26. Yang, B.; Jin, S.; Guo, S.; Park, Y.; Chen, L.; Zhao, B.; Jung, Y.M. Recent Development of SERS Technology: Semiconductor-Based Study. *ACS Omega* **2019**, *4*, 20101–20108. [CrossRef] [PubMed]
27. Perala, R.S.; Chandrasekar, N.; Balaji, R.; Alexander, P.S.; Humaidi, N.Z.N.; Hwang, M.T. A comprehensive review on graphene-based materials: From synthesis to contemporary sensor applications. *Mater. Sci. Eng. R Rep.* **2024**, *159*, 100805. [CrossRef]
28. Khyzhun, O.; Solonin, Y.; Dobrovolsky, V. Electronic structure of hexagonal tungsten trioxide: XPS, XES, and XAS studies. *J. Alloys Compd.* **2001**, *320*, 1–6. [CrossRef]
29. Zou, J.-W.; Li, Z.-D.; Kang, H.-S.; Zhao, W.-Q.; Liu, J.-C.; Chen, Y.-L.; Ma, L.; Hou, H.-Y.; Ding, S.-J. Strong Visible Light Absorption and Abundant Hotspots in Au-Decorated WO_3 Nanobricks for Efficient SERS and Photocatalysis. *ACS Omega* **2021**, *6*, 28347–28355. [CrossRef]
30. Yu, J.; Chen, C.; Zhang, Q.; Lin, J.; Yang, X.; Gu, L.; Zhang, H.; Liu, Z.; Wang, Y.; Zhang, S.; et al. Au Atoms Anchored on Amorphous C_3N_4 for Single-Site Raman Enhancement. *J. Am. Chem. Soc.* **2022**, *144*, 21908–21915. [CrossRef]
31. Chen, B.; Fan, L.; Li, C.; Xia, L.; Wang, K.; Wang, J.; Pang, D.; Zhu, Z.; Ma, P. Au nanoparticles decorated β-Bi_2O_3 as highly-sensitive SERS substrate for detection of methylene blue and methyl orange. *Analyst* **2024**, *149*, 4283–4294. [CrossRef] [PubMed]

32. Xu, L.; Zhang, H.; Tian, Y.; Jiao, A.; Chen, F.; Chen, M. Photochemical synthesis of ZnO@Au nanorods as an advanced reusable SERS substrate for ultrasensitive detection of light-resistant organic pollutant in wastewater. *Talanta* **2019**, *194*, 680–688. [CrossRef]
33. Huang, J.; Ma, D.; Chen, F.; Chen, D.; Bai, M.; Xu, K.; Zhao, Y. Green in Situ Synthesis of Clean 3D Chestnutlike Ag/WO_{3-x} Nanostructures for Highly Efficient, Recyclable and Sensitive SERS Sensing. *ACS Appl. Mater. Interfaces* **2017**, *9*, 7436–7446. [CrossRef]
34. Zhai, Y.; DuChene, J.S.; Wang, Y.-C.; Qiu, J.; Johnston-Peck, A.C.; You, B.; Guo, W.; DiCiaccio, B.; Qian, K.; Zhao, E.W.; et al. Polyvinylpyrrolidone-induced anisotropic growth of gold nanoprisms in plasmon-driven synthesis. *Nat. Mater.* **2016**, *15*, 889–895. [CrossRef]
35. Uboldi, C.; Bonacchi, D.; Lorenzi, G.; Hermanns, M.I.; Pohl, C.; Baldi, G.; E Unger, R.; Kirkpatrick, C.J. Gold nanoparticles induce cytotoxicity in the alveolar type-II cell lines A549 and NCIH441. *Part. Fibre Toxicol.* **2009**, *6*, 18. [CrossRef]
36. Lopez-Sanchez, J.A.; Dimitratos, N.; Hammond, C.; Brett, G.L.; Kesavan, L.; White, S.; Miedziak, P.; Tiruvalam, R.; Jenkins, R.L.; Carley, A.F.; et al. Facile removal of stabilizer-ligands from supported gold nanoparticles. *Nat. Chem.* **2011**, *3*, 551–556. [CrossRef]
37. Amendola, V.; Litti, L.; Meneghetti, M. LDI-MS Assisted by Chemical-Free Gold Nanoparticles: Enhanced Sensitivity and Reduced Background in the Low-Mass Region. *Anal. Chem.* **2013**, *85*, 11747–11754. [CrossRef]
38. Ma, S.; Xue, J.; Zhou, Y.; Zhang, Z. Photochemical synthesis of ZnO/Ag_2O heterostructures with enhanced ultraviolet and visible photocatalytic activity. *J. Mater. Chem. A* **2014**, *2*, 7272–7280. [CrossRef]
39. Ameer, F.S.; Pittman, C.U.; Zhang, D. Quantification of Resonance Raman Enhancement Factors for Rhodamine 6G (R6G) in Water and on Gold and Silver Nanoparticles: Implications for Single-Molecule R6G SERS. *J. Phys. Chem. C* **2013**, *117*, 27096–27104. [CrossRef]
40. Zheng, Q.; Fu, Y.-C.; Xu, J.-Q. Advances in the chemical sensors for the detection of DMMP—A simulant for nerve agent sarin. *Procedia Eng.* **2010**, *7*, 179–184. [CrossRef]
41. Bo, X.; Tang, A.; Dou, M.; Li, Z.; Wang, F. Controllable electrodeposition and mechanism research of nanostructured Bi_2Te_3 thin films with high thermoelectric properties. *Appl. Surf. Sci.* **2019**, *486*, 65–71. [CrossRef]
42. Lin, N.; Lin, Y.; Qian, X.; Wang, X.; Su, W. Construction of a 2D/2D $WO_3/LaTiO_2N$ Direct Z-Scheme Photocatalyst for Enhanced CO_2 Reduction Performance Under Visible Light. *ACS Sustain. Chem. Eng.* **2021**, *9*, 13686–13694. [CrossRef]
43. Dokou, E.; Stangland, E.E.; Andres, R.P.; Delgass, W.N.; Barteau, M.A. Comparison of AFM and HRTEM to determine the metal particle morphology and loading of an Au/TiO_2 catalyst. *Catal. Lett.* **2000**, *70*, 1–7. [CrossRef]
44. Sahu, B.K.; Dwivedi, A.; Pal, K.K.; Pandian, R.; Dhara, S.; Das, A. Optimized Au NRs for efficient SERS and SERRS performances with molecular and longitudinal surface plasmon resonance. *Appl. Surf. Sci.* **2021**, *537*, 147615. [CrossRef]
45. Kim, N.H.; Kim, K. Surface-enhanced resonance Raman scattering of rhodamine 6G on Pt nanoaggregates. *J. Raman Spectrosc.* **2005**, *36*, 623–628. [CrossRef]
46. Taranenko, N.; Alarie, J.-P.; Stokes, D.L.; Vo-Dinh, T. Surface-Enhanced Raman Detection of Nerve Agent Simulant (DMMP and DIMP) Vapor on Electrochemically Prepared Silver Oxide Substrates. *J. Raman Spectrosc.* **1996**, *27*, 379–384. [CrossRef]
47. Kim, S.; Park, S.; Park, S.; Lee, C. Acetone sensing of Au and Pd-decorated WO_3 nanorod sensors. *Sens. Actuators B Chem.* **2015**, *209*, 180–185. [CrossRef]
48. Foster, A.S.; Gejo, F.L.; Shluger, A.L.; Nieminen, R.M. Vacancy and interstitial defects in hafnia. *Phys. Rev. B* **2002**, *65*, 174117. [CrossRef]
49. Cong, S.; Yuan, Y.; Chen, Z.; Hou, J.; Yang, M.; Su, Y.; Zhang, Y.; Li, L.; Li, Q.; Geng, F.; et al. Noble metal-comparable SERS enhancement from semiconducting metal oxides by making oxygen vacancies. *Nat. Commun.* **2015**, *6*, 7800. [CrossRef]
50. Liu, H.; Chen, L.; Li, B.; Song, H.; Tan, C.L.; Shi, Y.; Yan, S. Semiconducting Tungsten Trioxide Thin Films for High-Performance SERS Biosensors. *Nanomaterials* **2025**, *15*, 1393. [CrossRef]
51. McMillan, K.S.; McCluskey, A.G.; Sorensen, A.; Boyd, M.; Zagnoni, M. Emulsion technologies for multicellular tumour spheroid radiation assays. *Analyst* **2015**, *141*, 100–110. [CrossRef]
52. Liang, J.; Zhang, L.; Wang, S.; Yu, Y.; Lei, D. Oxygen Vacancies Enhance SERS Performance of Tungsten-Doped Vanadium Dioxide Nanoparticles. *Adv. Mater. Technol.* **2024**, *10*, 2401304. [CrossRef]
53. Sharma, S.; Kumar, R.; Yadav, R.M. Polyacrylonitrile as a versatile matrix for gold nanoparticle-based SERS substrates. *Nanoscale Adv.* **2024**, *6*, 1065–1073. [CrossRef]
54. Yu, J.; Chen, C.; Lin, J.; Meng, X.; Qiu, L.; Wang, X. Amorphous $Co(OH)_2$ nanocages achieving efficient photo-induced charge transfer for significant SERS activity. *J. Mater. Chem. C* **2022**, *10*, 1632–1637. [CrossRef]
55. Wei, W.; Yao, Y.; Zhao, Q.; Xu, Z.; Wang, Q.; Zhang, Z.; Gao, Y. Oxygen defect-induced localized surface plasmon resonance at the WO_{3-x} quantum dot/silver nanowire interface: SERS and photocatalysis. *Nanoscale* **2019**, *11*, 5535–5547. [CrossRef]
56. Sneed, B.T.; Young, A.P.; Tsung, C.-K. Building up strain in colloidal metal nanoparticle catalysts. *Nanoscale* **2015**, *7*, 12248–12265. [CrossRef]
57. Ji, X.; Song, X.; Li, J.; Bai, Y.; Yang, W.; Peng, X. Size Control of Gold Nanocrystals in Citrate Reduction: The Third Role of Citrate. *J. Am. Chem. Soc.* **2007**, *129*, 13939–13948. [CrossRef] [PubMed]

58. Liu, W.; Bai, H.; Li, X.; Li, W.; Zhai, J.; Li, J.; Xi, G. Improved Surface-Enhanced Raman Spectroscopy Sensitivity on Metallic Tungsten Oxide by the Synergistic Effect of Surface Plasmon Resonance Coupling and Charge Transfer. *J. Phys. Chem. Lett.* **2018**, *9*, 4096–4100. [CrossRef]
59. Sun, J.; Hu, H.; Zheng, D.; Zhang, D.; Deng, Q.; Zhang, S.; Xu, H. Light-Emitting Plexciton: Exploiting Plasmon–Exciton Interaction in the Intermediate Coupling Regime. *ACS Nano* **2018**, *12*, 10393–10402. [CrossRef]
60. Yang, S.; Yao, J.; Quan, Y.; Hu, M.; Su, R.; Gao, M.; Han, D.; Yang, J. Monitoring the charge-transfer process in a Nd-doped semiconductor based on photoluminescence and SERS technology. *Light Sci. Appl.* **2020**, *9*, 117. [CrossRef]
61. Wang, X.-J.; Tian, X.; Sun, Y.-J.; Zhu, J.-Y.; Li, F.-T.; Mu, H.-Y.; Zhao, J. Enhanced Schottky effect of a 2D–2D CoP/g-C_3N_4 interface for boosting photocatalytic H_2 evolution. *Nanoscale* **2018**, *10*, 12315–12321. [CrossRef] [PubMed]
62. Handoko, A.D.; Steinmann, S.N.; Seh, Z.W. Theory-guided materials design: Two-dimensional MXenes in electro- and photocatalysis. *Nanoscale Horiz.* **2019**, *4*, 809–827. [CrossRef]
63. Hung, S.-F.; Xu, A.; Wang, X.; Li, F.; Hsu, S.-H.; Li, Y.; Wicks, J.; Cervantes, E.G.; Rasouli, A.S.; Li, Y.C.; et al. A Metal-Supported Single-Atom Catalytic Site Enables Carbon Dioxide Hydrogenation. *Nat. Commun.* **2022**, *13*, 819. [CrossRef]
64. Lei, Z.; Zhang, X.; Zhao, Y.; Wei, A.; Tao, L.; Yang, Y.; Zheng, Z.; Tao, L.; Yu, P.; Li, J. Enhanced Raman Scattering on Two-Dimensional Palladium Diselenide. *Nanoscale* **2022**, *14*, 4181–4187. [CrossRef]
65. Yang, L.; Peng, Y.; Yang, Y.; Liu, J.; Li, Z.; Ma, Y.; Zhang, Z.; Wei, Y.; Li, S.; Huang, Z.; et al. Green and Sensitive Flexible Semiconductor SERS Substrates: Hydrogenated Black TiO_2 Nanowires. *ACS Appl. Nano Mater.* **2018**, *1*, 4516–4527. [CrossRef]
66. Lin, J.; Shang, Y.; Li, X.; Yu, J.; Wang, X.; Guo, L. Ultrasensitive SERS Detection by Defect Engineering on Single Cu_2O Superstructure Particle. *Adv. Mater.* **2017**, *29*, 1604797. [CrossRef]
67. Zheng, X.; Zhong, H.; Wang, Z.; Li, J.; Hu, Y.; Li, H.; Jia, J.; Zhang, S.; Ren, F. Fabrication of Stable Substoichiometric WOx Films with High SERS Sensitivity by Thermal Treatment. *Vacuum* **2022**, *198*, 110884. [CrossRef]
68. Sun, Z.; Gao, Y.; Ban, C.; Meng, J.; Wang, J.; Wang, K.; Gan, L. 3 Nm-Wide WO_{3-x} Nanorods with Abundant Oxygen Vacancies as Substrates for High-Sensitivity SERS Detection. *ACS Appl. Nano Mater.* **2023**, *6*, 8635–8642. [CrossRef]

colloids and interfaces

MDPI

Article

Atomically Dispersed Pt–Sn Nanocluster Catalysts for Enhanced Toluene Hydrogenation in LOHC Systems

Jun Wang [1], Hao Lin [1], Qizhong Chan [2], Yaohong Zhao [1,*] and Xiaohui He [3,*]

1 National Institute of Guangdong Advanced Energy Storage Co., Ltd., Guangzhou 510080, China; wangjun190203@163.com (J.W.); haolam0616@163.com (H.L.)
2 Zhongshan Power Supply Bureau, Guangdong Power Grid Co., Ltd., Zhongshan 528405, China; chanwenxiang@126.com
3 Key Laboratory of Bioinorganic and Synthetic Chemistry of Ministry of Education, Fine Chemical Industry Research Institute, School of Chemistry, Institute of Green Chemistry and Molecular Engineering, Sun Yat-sen University, Guangzhou 510275, China
* Correspondence: kennyaoo@126.com (Y.Z.); hexiaohui@mail.sysu.edu.cn (X.H.)

Abstract

Liquid organic hydrogen carriers (LOHCs) are promising materials for safe, reversible, and high-density hydrogen storage. Atomically dispersed bimetallic Pt–Sn nanocluster catalysts supported on TiO_2 (Pt–Sn/TiO_2) were developed to enhance the hydrogenation step in the toluene-methylcyclohexane cycle, a model LOHC system. Compared with monometallic Pt/TiO_2 and Sn/TiO_2, Pt–Sn/TiO_2 exhibited superior hydrogenation performance. Mechanistic studies, including X-ray photoelectron spectroscopy, kinetic analysis, and H_2-D_2 exchange experiments, revealed that Sn incorporation modulates the electronic structure of Pt, enhancing H_2 activation and spillover. These findings provide insights into the rational design of atomically dispersed bimetallic nanocluster catalysts for efficient and durable hydrogen storage in LOHC-based systems.

Keywords: liquid organic hydrogen carriers; toluene hydrogenation; Pt; Sn; atomically dispersed nanocluster; electronic metal-support interaction

1. Introduction

Hydrogen energy has been recognized as one of the cleanest and most sustainable energy carriers due to its high gravimetric energy density, abundance, and environmentally benign combustion products [1–3]. The only by-product of hydrogen utilization is water, without the emission of greenhouse gases or pollutants, making hydrogen an ideal alternative to fossil fuels for achieving carbon neutrality [4,5]. In addition to its potential as a clean energy vector, hydrogen also serves as a critical feedstock in chemical, petrochemical, and metallurgical industries [6].

A complete hydrogen energy system generally consists of three essential components: hydrogen production, storage and transportation, and utilization [7]. Among these, hydrogen storage and transportation technologies remain key challenges for the large-scale deployment of hydrogen energy. Conventional storage methods, such as high-pressure gaseous storage and cryogenic liquid storage, are technologically mature but face issues including low volumetric density, high energy consumption, and potential safety risks such as leakage and material embrittlement [8,9]. These limitations highlight the urgent need for safer, more efficient, and more compact hydrogen storage technologies.

Liquid organic hydrogen carrier (LOHC) systems have emerged as a promising alternative, relying on the reversible hydrogenation and dehydrogenation of unsaturated organic molecules such as aromatics or olefins [10–14]. On 30 July 2021, Chiyoda Corporation and Mitsubishi Corporation jointly initiated a commercial-scale hydrogen import project based on LOHC technology, while companies in the Netherlands and Japan have also explored the use of LOHC systems for hydrogen transport. Compared with high-pressure or cryogenic storage, LOHC technology offers several advantages, including higher storage density, fully reversible hydrogenation/dehydrogenation cycles, simpler handling and transport, and improved safety [15–17]. Newson et al. evaluated various liquid organic hydrogen carriers and identified aromatic hydrocarbons as the most promising candidates, owing to their high storage capacity, favorable thermodynamics, and excellent reversibility during hydrogenation-dehydrogenation cycles [18]. Among these, the toluene-methylcyclohexane-hydrogen (MTH) cycle has been extensively investigated as a model LOHC system [19]. The hydrogenation of toluene to methylcyclohexane represents the hydrogen storage step in this reversible cycle and plays a crucial role in determining the overall system efficiency. Due to the inherent thermodynamic and kinetic stability of the aromatic ring, practical toluene hydrogenation often requires relatively high temperatures and elevated hydrogen pressures to reach desirable conversion levels [20]. Therefore, developing efficient and durable catalysts for toluene hydrogenation is essential for advancing LOHC-based hydrogen storage technologies.

Noble metal catalysts (e.g., Pt, Pd, Ru) have demonstrated excellent hydrogen activation capability and high catalytic activity in toluene hydrogenation [21,22]. Nevertheless, their high cost and limited natural abundance hinder large-scale application. To address these challenges, atomically dispersed metal catalysts have attracted growing attention due to their nearly 100% atomic utilization efficiency and tunable electronic properties [23–25]. Such catalysts—ranging from single-atom to dual-atom configurations—enable precise control of active sites and reaction pathways [26]. The introduction of a secondary metal can further modulate the electronic structure of the active center, enhance hydrogen adsorption and dissociation, and suppress undesired side reactions.

In this study, Pt–Sn/TiO_2 atomically dispersed nanocluster catalysts were synthesized via an impregnation method and comprehensively characterized to establish the correlation between structure and catalytic performance. The catalysts were evaluated in toluene hydrogenation, the key hydrogen storage step in the MTH cycle, to assess their potential for LOHC applications.

2. Materials and Methods

2.1. Chemicals

All reagents were used as received without further purification. Chloroplatinic acid hexahydrate ($H_2PtCl_6 \cdot 6H_2O$, 99.9%), nano-TiO_2 (P25, 99%), tin(IV) chloride pentahydrate ($SnCl_4 \cdot 5H_2O$, $\geq$99.9%), methanol (99.8%) and tridecane (>99%) were purchased from Shanghai Macklin Biochemical Co., Ltd. (Shanghai, China). Toluene (99.8%) was purchased from Shanghai Anpu Experimental Technology Co., Ltd. (Shanghai, China). Methylcyclohexane (99%), anhydrous ethanol ($\geq$99%), cyclohexane (chromatographic grade), and isopropanol (98%) were purchased from Shanghai Aladdin Biochemical Co., Ltd. (Shanghai, China). Hydrogen (99%) was purchased from Guangzhou Guangqi Gas Co., Ltd. (Shanghai, China).

2.2. Apparatus

Catalyst synthesis and characterization were conducted using standard laboratory equipment, including a muffle furnace (KSL-1750X, Hefei Kejing Material Technology Co., Ltd., Hefei, China), a magnetic stirrer bath (HWCL-1, Zhengzhou Great Wall Science &

Trade Co., Ltd., Zhengzhou, China), a low-speed peristaltic pump (LHZW006, United Zhongwei Technology Co., Ltd., Zhengzhou, China), an analytical balance (ML204, Mettler-Toledo, Columbus, OH, USA), and a high-speed centrifuge (TG16-WT, Shenzhen Vector Scientific Instruments Co., Ltd., Shenzhen, China).

Structural and surface characterizations were performed using X-ray powder diffraction (XRD, D-MAX 2200 VPC, Rigaku Corporation, Tokyo, Japan), X-ray photoelectron spectroscopy (XPS, Escalab 250XPS, Thermo Fisher Scientific, Waltham, MA, USA), inductively coupled plasma optical emission spectroscopy (ICP-OES, PerkinElmer Optima 8000, PerkinElmer Inc., Waltham, MA, USA), BET surface area analysis (ASAP2460, Micromeritics, Norcross, GA, USA), transmission electron microscopy (TEM, FEI Tecnai G2 F30, FEI, Hillsboro, OR, USA), and aberration-corrected high-angle annular dark-field scanning TEM (AC HAADF-STEM, JEM-ARM200P, JEOL Ltd., Tokyo, Japan).

Product analysis was performed with gas chromatography (GC2010 plus, Shimadzu, Kyoto, Japan) and GC-MS (GCMS-QP2010-Ultra, Shimadzu, Kyoto, Japan). The H_2-D_2 exchange reaction was conducted in a continuous-flow fixed-bed quartz reactor with an internal diameter of 7 mm under ambient temperature and atmospheric pressure. About 100 mg of the catalyst sample was loaded into the reactor and held in place with quartz cotton plugs. A mixture of 5% H_2 in N_2 was used as the carrier gas at a flow rate of 20 mL/min, while D_2 was introduced as a pulse gas flowing at 10 mL/min. The signal of HD (m/z = 3) was monitored in real time using a mass spectrometer (Hiden Analytical HPR-20 QIC benchtop gas analysis system, Warrington, UK).

2.3. *Catalyst Preparation*

2.3.1. Pt/TiO_2 Atomically Dispersed Nanocluster Catalyst

Chloroplatinic acid hexahydrate ($H_2PtCl_6{\cdot}6H_2O$, 0.63 g) was accurately weighed and dissolved in 5 mL of deionized water to prepare a 0.1 g/mL $H_2PtCl_6{\cdot}$solution. An aliquot of 210 μL of this solution was diluted with 100 mL of deionized water. Separately, nano-TiO_2 (P25, 1 g) was dispersed in 150 mL of deionized water and stirred at 800 rpm for 300 min. During stirring, the diluted H_2PtCl_6 solution was added dropwise to the TiO_2 suspension at a rate of 1.42 mL/min using a low-speed peristaltic pump. The resulting mixture was centrifuged, and the collected solid was dried overnight at 60 °C, followed by calcination in air at 400 °C for 2 h with a heating rate of 5 °C/min to obtain the Pt/TiO_2 catalyst.

2.3.2. Sn/TiO_2 Atomically Dispersed Nanocluster Catalyst

Tin(IV) chloride pentahydrate ($SnCl_4{\cdot}5H_2O$, 29.5 mg) was dissolved in 100 mL of deionized water. Meanwhile, nano-TiO_2 (P25, 1 g) was dispersed in 150 mL deionized water and stirred at 800 rpm for 300 min. During stirring, the $SnCl_4$ solution was added dropwise to the TiO_2 suspension at a rate of 1.42 mL/min using a low-speed peristaltic pump. Subsequent centrifugation, drying, and calcination steps were identical to those described for Pt/TiO_2.

2.3.3. Pt–Sn/TiO_2 Atomically Dispersed Nanocluster Catalyst

Chloroplatinic acid hexahydrate ($H_2PtCl_6{\cdot}6H_2O$, 0.63 g) was accurately weighed and dissolved in 5 mL of deionized water to prepare a 0.1 g/mL H_2PtCl_6 solution. An aliquot of 210 μL of this solution and tin(IV) chloride pentahydrate ($SnCl_4{\cdot}5H_2O$, 3.7 mg) were added to 100 mL of deionized water. Nano-TiO_2 (P25, 1 g) was dispersed in 150 mL of deionized water and stirred at 800 rpm for 300 min. During stirring, the mixed metal precursor solution was added dropwise to the TiO_2 suspension at a rate of 1.42 mL/min using a low-speed peristaltic pump. The subsequent procedures were identical to those described above for the other catalysts.

2.4. Catalyst Performance Analysis Methods

2.4.1. Toluene Hydrogenation

The catalytic hydrogenation of toluene was conducted in a 10 mL stainless-steel autoclave. In a typical experiment, 20 mg of catalyst, 0.5 mmol of toluene, and 0.12 mmol of tridecane (internal standard) were dissolved in 2 mL of cyclohexane. The reactor was purged with H_2 five times before being pressurized to 1 MPa H_2. The reaction was carried out at 100 °C under magnetic stirring. After completion, the reaction mixture was filtered, and the liquid products were analyzed by GC and GC-MS.

2.4.2. Catalyst Stability

The used Pt–Sn/TiO_2 catalyst was recovered by filtration, washed with ethanol and deionized water, dried at 60 °C for 10 h, and re-calcined at 400 °C for 2 h. The regenerated catalyst was reused under identical reaction conditions for stability tests.

2.4.3. Data Analysis

Toluene conversion and product selectivity were calculated using the following equations:

$$\text{Conversion}(\%) = \frac{C_0 - C_1}{C_0} \quad (1)$$

$$\text{Selectivity}(\%) = \frac{C_2}{C_0 - C_1} \quad (2)$$

where C_0 is the initial molar amount of toluene, C_1 is the molar amount of unreacted toluene, and C_2 is the molar amount of the product.

The activity of catalyst was calculated as:

$$\text{Activity}\left(h^{-1}\right) = \frac{\text{mols of substrate consumed}}{\text{mols of active metal (mmol)} \times \text{reaction time}} \quad (3)$$

3. Results and Discussion

3.1. Catalyst Characterization

Efficient hydrogenation catalysts are critical not only for fundamental studies of reaction mechanisms, but also for applications in organic liquid hydrogen carriers (LOHCs) for reversible hydrogen storage. In this work, Pt–Sn/TiO_2 catalysts were synthesized by impregnating TiO_2 with chloroplatinic acid and tin(IV) chloride pentahydrate, followed by calcination at 400 °C for 2 h (Figure S1). Pt/TiO_2 and Sn/TiO_2 catalysts were prepared under identical conditions for comparison. The catalysts were characterized to evaluate their structural and morphological features. Metal loadings were determined by inductively coupled plasma optical emission spectroscopy (ICP-OES), as summarized in Table S1. Pt–Sn/TiO_2 contained 0.68 wt% Pt and 0.089 wt% Sn, whereas Pt/TiO_2 and Sn/TiO_2 contained 0.72 wt% Pt and 0.83 wt% Sn, respectively. The observed loadings were in good agreement with the theoretical values, indicating that the impregnation method is an effective approach for preparing supported catalysts.

The crystalline structures of the catalysts and the TiO_2 support were analyzed by X-ray diffraction (XRD), as shown in Figure 1a. All samples exhibited diffraction patterns consistent with the anatase phase of TiO_2, indicating that metal loading did not alter the crystal structure. No characteristic peaks of metallic Pt (PDF#04-0802) or Sn (PDF#18-1380) were detected in Pt/TiO_2, Sn/TiO_2, or Pt–Sn/TiO_2, implying that both Pt and Sn species are highly dispersed on the TiO_2 surface. The absence of metal peaks may also result from the low metal contents, which are below the detection limit of the instrument.

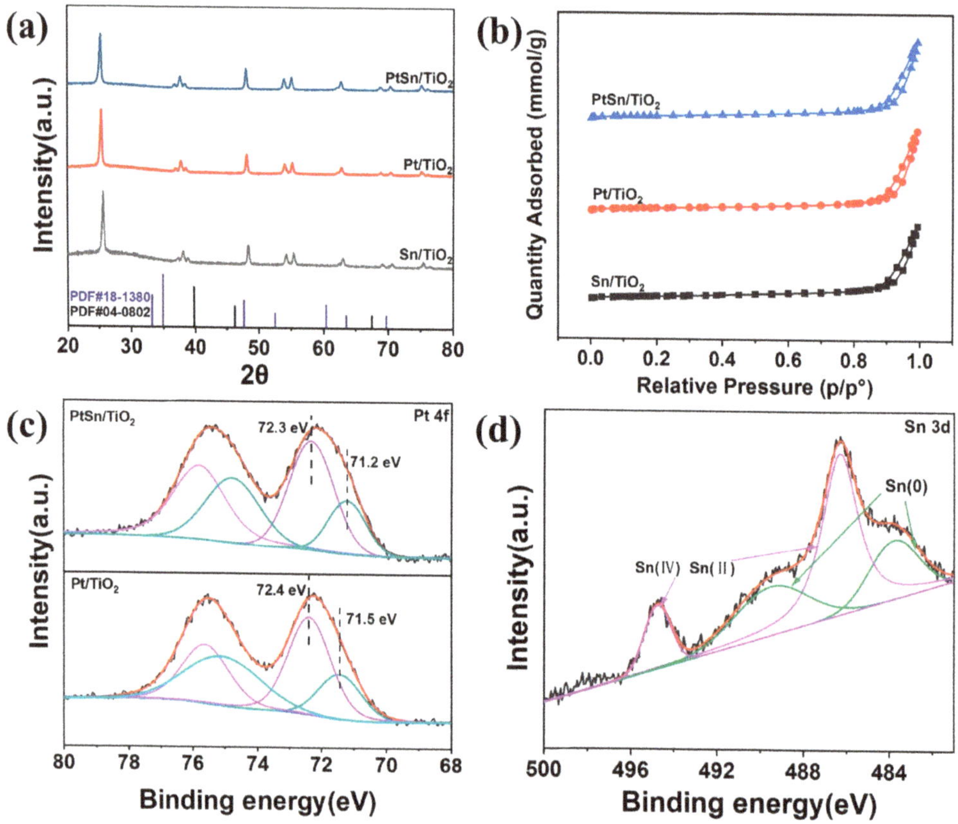

Figure 1. (**a**) XRD image, (**b**) N_2 adsorption/desorption isotherm of TiO_2, Sn/TiO_2, Pt/TiO_2 and $Pt–Sn/TiO_2$, (**c**) XPS of Pt 4f of $Pt–Sn/TiO_2$ and Pt/TiO_2, (**d**) XPS of Sn 3d of $Pt–Sn/TiO_2$.

The textural properties were further investigated by N_2 physisorption (Figure 2b). All catalysts exhibited Type IV isotherms, characteristic of mesoporous materials. The BET surface areas of Pt/TiO_2, $Pt–Sn/TiO_2$, and Sn/TiO_2 were 38.5, 38.9, and 39.2 $m^2 \cdot g^{-1}$, respectively (Table S2), exhibiting minor differences. The comparable pore volumes and average pore sizes across the series confirm that the mesoporous structure of the TiO_2 support was well-preserved after metal loading. The BET results, showing preserved mesoporosity, are consistent with the XRD analysis, indicating that metal loading did not significantly affect the TiO_2 framework.

To further investigate the spatial distribution of metal species, transmission electron microscopy (TEM) and aberration-corrected scanning transmission electron microscopy (AC-STEM) were performed. Representative TEM, STEM, AC-STEM, and energy-dispersive X-ray spectroscopy (EDS) mapping images of $Pt–Sn/TiO_2$, Pt/TiO_2, and Sn/TiO_2 are presented in Figures 2, S2 and S3. The dark-field STEM images (Figures 2b, S2b and S3b) display uniformly distributed bright spots at a 20 nm scale, indicating a homogeneous dispersion of metallic species. High-resolution AC-STEM images (Figures 2c, S2c and S3c) further reveal that the metals are present as uniformly dispersed nanoclusters on the TiO_2 surface. The corresponding EDS mapping (Figures 2d, S2d and S3d) further supports these observations, showing that Pt and Sn are distributed over the support. These results demonstrate the successful formation of $Pt–Sn/TiO_2$, Pt/TiO_2, and Sn/TiO_2 catalysts with atomically dispersed nanoclusters via the impregnation method.

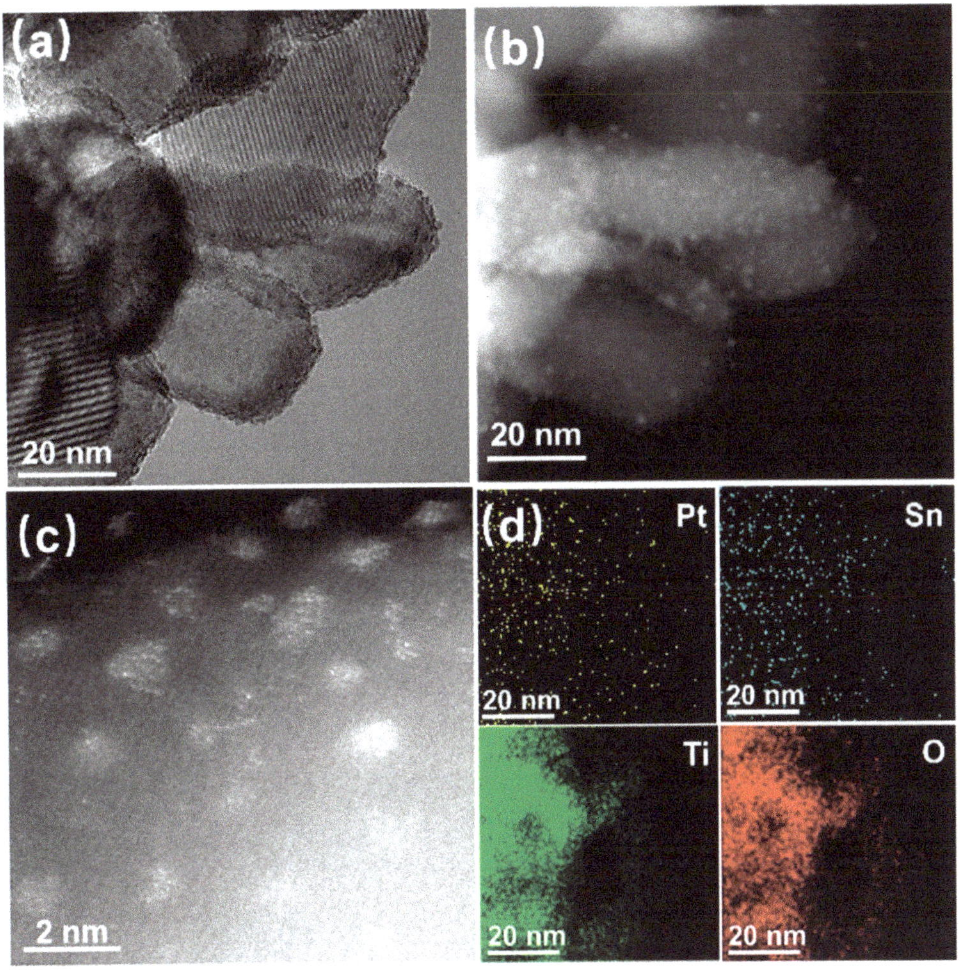

Figure 2. (**a**) TEM image, (**b**) STEM image, (**c**) AC HADDF-STEM image and (**d**) EDS mapping image of Pt–Sn/TiO_2.

X-ray photoelectron spectroscopy (XPS) was employed to investigate the oxidation states of Pt and Sn in Pt/TiO_2 and Pt–Sn/TiO_2 catalysts. The Pt 4f spectra of Pt/TiO_2 and Pt–Sn/TiO_2 (Figure 1c) exhibits two peaks at 71.5 eV and 72.4 eV, assigned to Pt^0 $4f_{7/2}$ and Pt^{2+} $4f_{7/2}$ in Pt/TiO_2, indicating the coexistence of metallic and oxidized Pt species. After introducing Sn, the Pt 4f peaks in Pt–Sn/TiO_2 shift slightly to lower binding energies (71.2 eV and 72.3 eV) without changing the valence states. This shift is attributed to electron transfer from Sn to the more electronegative Pt, which increases the electron density on Pt and generates an electron-rich state. The Sn 3d spectrum of Pt–Sn/TiO_2 (Figure 1d) exhibits four peaks: 483.6 eV and 489.1 eV for Sn^0, and 486.3 eV and 494.7 eV for oxidized Sn species (Sn^{2+} and Sn^{4+}) [27,28]. These findings indicate the coexistence of metallic and oxidized Pt and Sn, and demonstrate electronic interactions between the two metals in the bimetallic catalyst.

3.2. Catalytic Performance Evaluation

Toluene hydrogenation, representing the hydrogen storage step in the toluene-methylcyclohexane (MTH) cycle for LOHC systems, was employed to evaluate the catalytic performance of Pt–Sn/TiO_2 and Pt/TiO_2. The effect of solvent was tested at 100 °C and 1 MPa H_2 for 55 min in cyclohexane, ethanol, and isopropanol (IPA). Pt–Sn/TiO_2 showed the highest toluene conversion in cyclohexane (92.1%), significantly higher than in ethanol (42.1%) and IPA (55.2%) (Figure S4). Cyclohexane was thus selected as the reaction solvent for all subsequent experiments. Toluene conversion over Pt–Sn/TiO_2 and Pt/TiO_2 increased with reaction time, reaching >99.9% and 81.3% after 65 min, respectively, while Sn/TiO_2 was almost inactive (<1%) (Figure 3a). These results indicate that metallic Pt acts as the active center for toluene hydrogenation, and the incorporation of Sn promotes the reaction. At conversions below 30% and 100 °C, the activity of Pt–Sn/TiO_2 was 1992.1 h^{-1}, 1.4 times higher than Pt/TiO_2 (1462.7 h^{-1}) (Figure 3b). Temperature-dependent tests under 1 MPa H_2 from 80 to 110 °C showed that Pt–Sn/TiO_2 consistently outperformed Pt/TiO_2 (Figure 3c). Pt–Sn/TiO_2 also showed excellent recyclability, maintaining >99.9% toluene conversion and 97.4% methylcyclohexane selectivity over five cycles (Figure 3d). STEM images before and after reaction (Figure S5a,b) revealed that Pt remained as well-dispersed nanoclusters with only minor aggregation (average size increased from 1.15 nm to 1.5 ± 0.5 nm), indicating good structural stability under reaction conditions. Such durability is crucial for LOHC systems, which require catalysts capable of enduring repeated hydrogenation–dehydrogenation cycles. The high activity and stability of Pt–Sn/TiO_2 suggest its potential as an efficient catalyst for LOHC hydrogen storage, laying the groundwork for further investigation under solvent-free and extended cycling conditions.

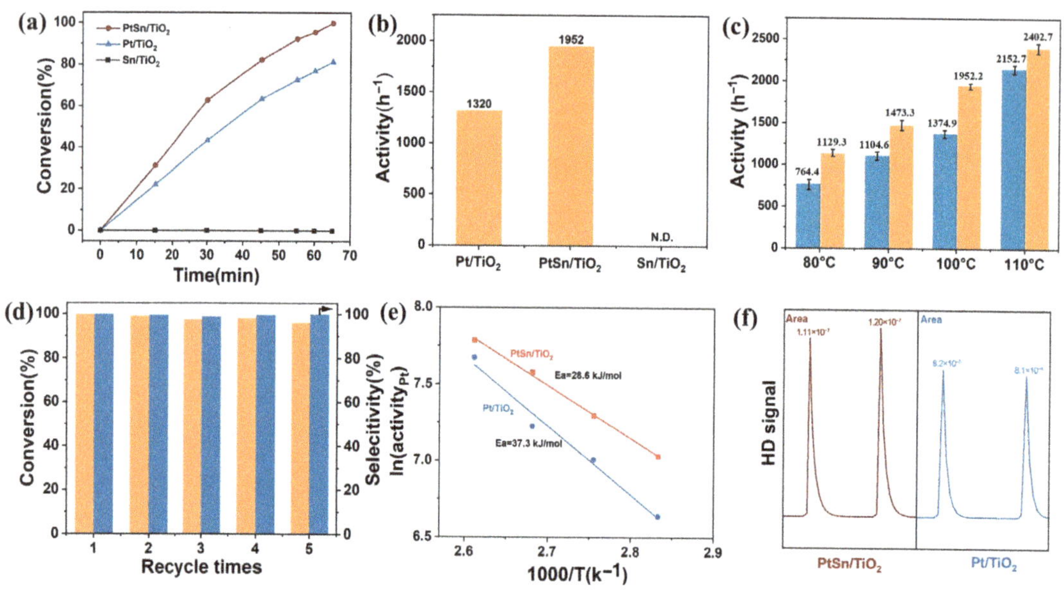

Figure 3. (**a**) Effect of reaction time on the catalytic performance for Pt–Sn/TiO_2, Pt/TiO_2 and Sn/TiO_2 (20 mg of catalyst, 0.5 mmol of toluene, T = 100 °C, 1 MPa H_2), (**b**) Reaction activity of Pt–Sn/TiO_2, Pt/TiO_2 and Sn/TiO_2 (Conversion below 30%), (**c**) Effect of temperature on the activity of Pt–Sn/TiO_2 and Pt/TiO_2 (20 mg of catalyst, 0.5 mmol of toluene, 1 MPa H_2), (**d**) Recycle performance of Pt–Sn/TiO_2 (Reaction conditions: 20 mg of Pt–Sn/TiO_2, 0.5 mmol of toluene, 2 mL of cyclohexane, T = 100 °C, 1 MPa H_2, time = 65 min), (**e**) Arrhenius plots and (**f**) H_2-D_2 exchange of Pt–Sn/TiO_2 and Pt/TiO_2.

The enhanced hydrogenation performance of Pt–Sn/TiO_2 highlights its potential for efficient hydrogen storage in LOHC systems, where rapid and reversible H_2 uptake is critical. To investigate the origin of the performance difference between Pt–Sn/TiO_2 and Pt/TiO_2 in toluene hydrogenation, the apparent activation energies were determined from Arrhenius plots (Figure 3e). Pt–Sn/TiO_2 exhibited a lower activation energy (28.6 kJ/mol) than Pt/TiO_2 (37.3 kJ/mol), indicating that Sn incorporation enhances the catalytic efficiency, consistent with the activity results shown in Figure 3b. The H_2 dissociation abilities of the two catalysts were further evaluated by H-D exchange experiments. The two peaks observed for each catalyst in Figure 3f correspond to sequential hydrogen activation events on different types of active sites, reflecting the catalyst's ability to dissociate molecular hydrogen. The HD peak area of Pt–Sn/TiO_2 was 1.5 times larger than that of Pt/TiO_2, demonstrating that the bimetallic Pt–Sn catalyst has significantly improved H_2 activation. These results confirm that Sn doping enhances the hydrogenation performance of Pt for toluene. Overall, Pt–Sn/TiO_2 shows higher activity in toluene hydrogenation than most catalysts reported in the literature [21,22,29–33]. This superior activity can be attributed to the highly dispersed architecture of the catalyst and the consequent Pt–Sn synergy, which collectively enhance the hydrogen dissociation capability.

4. Conclusions

In summary, Pt–Sn/TiO_2 atomically dispersed nanocluster catalysts were successfully synthesized via an impregnation method and thoroughly characterized. Structural analyses (XRD, XPS, TEM, and AC-STEM) confirmed uniform dispersion of Pt and Sn nanoclusters with evident electronic interactions. Pt–Sn/TiO_2 exhibited superior activity in toluene hydrogenation (activity = 1992.1 h^{-1}) compared with Pt/TiO_2 (1462.7 h^{-1}), along with excellent recyclability. Kinetic analysis revealed a lower apparent activation energy, while H-D exchange experiments demonstrated enhanced H_2 dissociation, accounting for its improved catalytic performance. The high activity, stability, and electronic synergy of Pt–Sn/TiO_2 highlight its potential as an efficient hydrogenation catalyst for organic liquid hydrogen carriers (LOHCs). These findings provide valuable insights for designing robust, interface-engineered catalysts to enable efficient hydrogen storage and release in sustainable energy systems.

Supplementary Materials: The following supporting information can be downloaded at: https://www.mdpi.com/article/10.3390/colloids9060085/s1, Figure S1. Schematic diagram of the preparation process for Pt/TiO_2, Sn/TiO_2 and Pt–Sn/TiO_2 atomically dispersed nanocluster catalysts; Table S1. Pt and Sn loading of Pt–Sn/TiO_2, Pt/TiO_2 and Sn/TiO_2; Table S2. The BET surface area, pore volume and pore size of different catalysts; Figure S2. (a) TEM image, (b) STEM image, (c) AC HADDF-STEM image and (d) EDS mapping image of Pt/TiO_2; Figure S3. (a) TEM image, (b) STEM image, (c) AC HADDF-STEM image and (d) EDS mapping image of Sn/TiO_2; Figure S4. Effect of different solvents on reaction properties (20 mg of Pt–Sn/TiO_2, 0.5 mmol of toluene, T = 100 °C, 1MPa H_2, time = 55 min); Figure S5. (a) STEM image and particle size distribution of Pt–Sn/TiO_2, (b) STEM image and particle size distribution of used Pt–Sn/TiO_2; Table S3. Comparison of the catalytic performance of Pt–Sn/TiO_2 with other catalysts.

Author Contributions: Conceptualization, J.W. and X.H.; methodology, J.W. and H.L.; validation, H.L. and Q.C.; formal analysis, J.W.; investigation, J.W. and H.L.; resources, Y.Z. and X.H.; data curation, J.W.; writing—original draft preparation, J.W. and X.H.; writing—review and editing, Y.Z. and X.H.; visualization, J.W.; supervision, Y.Z. and X.H.; project administration, Y.Z. and X.H.; funding acquisition, Y.Z. and X.H. All authors have read and agreed to the published version of the manuscript.

Funding: This work was supported by the National Natural Science Foundation of China (22422815, U22A20428), Guangdong Natural Science Funds for Distinguished Young Scholar (2022B1515020035), the special fund for Science and Technology Innovation Teams of Shanxi Province (202304051001007). The authors thank the Guangdong Basic Research Center of Excellence for Functional Molecular Engineering, and the Chemistry and Chemical Engineering Guangdong Laboratory (Grant No. 1922010).

Data Availability Statement: The raw data supporting the conclusions of this article will be made available by the authors on request.

Acknowledgments: We are indebted to our colleagues at the Guangdong Basic Research Center of Excellence for Functional Molecular Engineering for their valuable discussions and to the Chemistry and Chemical Engineering Guangdong Laboratory for providing access to characterization facilities. Additionally, we acknowledge the support from the National Institute of Guangdong Advanced Energy Storage Co., Ltd. and Zhongshan Power Supply Bureau.

Conflicts of Interest: Author Jun Wang, Hao Lin and Yaohong Zhao were employed by the company Guangdong Advanced Energy Storage Co., Ltd. Author Qizhong Chan was employed by the company Guangdong Power Grid Co., Ltd. The remaining authors declare that the research was conducted in the absence of any commercial or financial relationships that could be construed as a potential conflict of interest. All authors have read and agreed to the published version of the manuscript.

References

1. Brandon, N.P.; Kurban, Z. Clean Energy and the Hydrogen Economy. *Philos. Trans. R. Soc. A* **2017**, *375*, 20160400. [CrossRef]
2. Tarhan, C.; Çil, M.A. A Study on Hydrogen, the Clean Energy of the Future: Hydrogen Storage Methods. *J. Energy Storage* **2021**, *40*, 102676. [CrossRef]
3. Dong, Z.Y.; Yang, J.; Yu, L.; Daiyan, R.; Amal, R. A Green Hydrogen Credit Framework for International Green Hydrogen Trading Towards a Carbon Neutral Future. *Int. J. Hydrogen Energy* **2022**, *47*, 728–734. [CrossRef]
4. Lee, H.; Ahn, J.; Choi, D.G.; Park, S.Y. Analysis of the Role of Hydrogen Energy in Achieving Carbon Neutrality by 2050: A Case Study of the Republic of Korea. *Energy* **2024**, *304*, 132023. [CrossRef]
5. Shafiee, R.T.; Schrag, D.P. Carbon Abatement Costs of Green Hydrogen across End-Use Sectors. *Joule* **2024**, *8*, 3281–3289. [CrossRef]
6. Zhu, Y.; Keoleian, G.A.; Cooper, D.R. The Role of Hydrogen in Decarbonizing U.S. Industry: A Review. *Renew. Sustain. Energy Rev.* **2025**, *214*, 115392. [CrossRef]
7. Rampai, M.M.; Mtshali, C.B.; Seroka, N.S.; Khotseng, L. Hydrogen Production, Storage, and Transportation: Recent Advances. *RSC Adv.* **2024**, *14*, 6699–6718. [CrossRef]
8. Sarmah, M.K.; Singh, T.P.; Kalita, P.; Dewan, A. Sustainable Hydrogen Generation and Storage—A Review. *RSC Adv.* **2023**, *13*, 25253–25275. [CrossRef]
9. Zhang, T.; Uratani, J.; Huang, Y.; Xu, L.; Griffiths, S.; Ding, Y. Hydrogen Liquefaction and Storage: Recent Progress and Perspectives. *Renew. Sustain. Energy Rev.* **2023**, *176*, 113204. [CrossRef]
10. Tang, C.; Tang, S.; Sha, F.; Han, Z.; Feng, Z.; Wang, J.; Li, C. Insights into the Selectivity Determinant and Rate-Determining Step of CO_2 Hydrogenation to Methanol. *J. Phys. Chem. C* **2022**, *126*, 10399–10407. [CrossRef]
11. Peters, W.; Seidel, A.; Herzog, S.; Bösmann, A.; Schwieger, W.; Wasserscheid, P. Macrokinetic Effects in Perhydro-N-Ethylcarbazole Dehydrogenation and H_2 productivity Optimization by Using Egg-Shell Catalysts. *Energy Environ. Sci.* **2015**, *8*, 3013–3021. [CrossRef]
12. Modisha, P.M.; Ouma, C.N.M.; Garidzirai, R.; Wasserscheid, P.; Bessarabov, D. The Prospect of Hydrogen Storage Using Liquid Organic Hydrogen Carriers. *Energy Fuels* **2019**, *33*, 2778–2796. [CrossRef]
13. Le, T.-H.; Tran, N.; Lee, H.-J. Development of Liquid Organic Hydrogen Carriers for Hydrogen Storage and Transport. *Int. J. Mol. Sci.* **2024**, *25*, 1359. [CrossRef] [PubMed]
14. Snider, J.L.; Su, J.; Verma, P.; El Gabaly, F.; Sugar, J.D.; Chen, L.; Chames, J.M.; Talin, A.A.; Dun, C.; Urban, J.J.; et al. Stabilized Open Metal Sites in Bimetallic Metal-Organic Framework Catalysts for Hydrogen Production from Alcohols. *J. Mater. Chem. A* **2021**, *9*, 10869–10881. [CrossRef]
15. Modisha, P.; Bessarabov, D. Aromatic Liquid Organic Hydrogen Carriers for Hydrogen Storage and Release. *Curr. Opin. Green Sustain. Chem.* **2023**, *42*, 100820. [CrossRef]
16. Meda, U.S.; Raikar, O.M.; Acharya, A.; Kashyap S G, A.; Mahesh, R.; Shetty, T. Nuances in Liquid Organic Hydrogen Carriers. *Renew. Sustain. Energy Rev.* **2026**, *226*, 116237. [CrossRef]

17. Wu, Y.; Guo, Y.; Yu, H.; Jiang, X.; Zhang, Y.; Qi, Y.; Fu, K.; Xie, L.; Li, G.; Zheng, J.; et al. Nonstoichiometric Yttrium Hydride–Promoted Reversible Hydrogen Storage in a Liquid Organic Hydrogen Carrier. *CCS Chem.* **2021**, *3*, 974–984. [CrossRef]
18. Scherer, G. Economic Analysis of the Seasonal Storage of Electricity with Liquid Organic Hydrides. *Int. J. Hydrogen Energy* **1999**, *24*, 1157–1169. [CrossRef]
19. Kwak, Y.; Kirk, J.; Moon, S.; Ohm, T.; Lee, Y.-J.; Jang, M.; Park, L.-H.; Ahn, C.-I.; Jeong, H.; Sohn, H.; et al. Hydrogen Production from Homocyclic Liquid Organic Hydrogen Carriers (LOHCs): Benchmarking Studies and Energy-Economic Analyses. *Energy Convers. Manag.* **2021**, *239*, 114124. [CrossRef]
20. Li, J.; Zhang, Q.; Zhang, J.; Gao, R.; Zhang, X.; Zou, J.-J.; Pan, L. Intrinsic Kinetics of Dibenzyltoluene Hydrogenation over a Supported Ni Catalyst for Green Hydrogen Storage. *Ind. Eng. Chem. Res.* **2023**, *63*, 220–232. [CrossRef]
21. Yang, Y.; Lin, X.; Tang, J.; Zhang, J.; Liu, C.; Huang, J. Supported Mesoporous Pt Catalysts with Excellent Performance for Toluene Hydrogenation under Low Reaction Pressure. *Mol. Catal.* **2022**, *524*, 112341. [CrossRef]
22. Suppino, R.S.; Landers, R.; Cobo, A.J.G. Influence of Noble Metals (Pd, Pt) on the Performance of Ru/Al_2O_3 Based Catalysts for Toluene Hydrogenation in Liquid Phase. *Appl. Catal. A* **2016**, *525*, 41–49. [CrossRef]
23. He, X.; Zhang, H.; Zhang, X.; Zhang, Y.; He, Q.; Chen, H.; Cheng, Y.; Peng, M.; Qin, X.; Ji, H.; et al. Building up Libraries and Production Line for Single Atom Catalysts with Precursor-Atomization Strategy. *Nat. Commun.* **2022**, *13*, 5721. [CrossRef]
24. Cui, X.; Li, W.; Ryabchuk, P.; Junge, K.; Beller, M. Bridging Homogeneous and Heterogeneous Catalysis by Heterogeneous Single-Metal-Site Catalysts. *Nat. Catal.* **2018**, *1*, 385–397. [CrossRef]
25. Qiao, B.; Wang, A.; Yang, X.; Allard, L.F.; Jiang, Z.; Cui, Y.; Liu, J.; Li, J.; Zhang, T. Single-Atom Catalysis of Co Oxidation Using Pt_1/FeO_x. *Nat. Chem.* **2011**, *3*, 634–641. [CrossRef]
26. Liu, H.; Liu, L.; Qin, Q.; Li, J.; Li, B.; He, X.; Ji, H. Comparative Study of Ptm (M=Cu, Zn, Ga, Mn, Fe, in, Ce) Bimetals on Zincosilicate for Propane Dehydrogenation Reaction. *Chemistry* **2024**, *30*, e202402764. [CrossRef]
27. Liu, H.; Li, B.; Liu, Z.; Liang, Z.; Chuai, H.; Wang, H.; Lou, S.N.; Su, Y.; Zhang, S.; Ma, X. Ceria-Mediated Dynamic $Sn^0/Sn^{\delta+}$ Redox Cycle for CO_2 Electroreduction. *ACS Catal.* **2023**, *13*, 5033–5042. [CrossRef]
28. Xing, Y.; Kang, L.; Ma, J.; Jiang, Q.; Su, Y.; Zhang, S.; Xu, X.; Li, L.; Wang, A.; Liu, Z.-P.; et al. Sn_1pt Single-Atom Alloy Evolved Stable PtSn/Nano-Al_2O_3 Catalyst for Propane Dehydrogenation. *Chin. J. Catal.* **2023**, *48*, 164–174. [CrossRef]
29. Champness, N.R. The Future of Metal-Organic Frameworks. *Dalton Trans.* **2011**, *40*, 10311–10315. [CrossRef] [PubMed]
30. Chen, L.; Verma, P.; Hou, K.; Qi, Z.; Zhang, S.; Liu, Y.S.; Guo, J.; Stavila, V.; Allendorf, M.D.; Zheng, L.; et al. Reversible Dehydrogenation and Rehydrogenation of Cyclohexane and Methylcyclohexane by Single-Site Platinum Catalyst. *Nat. Commun.* **2022**, *13*, 1092. [CrossRef] [PubMed]
31. Park, K.H.; Jang, K.; Kim, H.J.; Son, S.U. Near-Monodisperse Tetrahedral Rhodium Nanoparticles on Charcoal: The Shape-Dependent Catalytic Hydrogenation of Arenes. *Angew. Chem. Int. Ed. Engl.* **2007**, *46*, 1152–1155. [CrossRef] [PubMed]
32. Stanley, J.N.G.; Heinroth, F.; Weber, C.C.; Masters, A.F.; Maschmeyer, T. Robust Bimetallic Pt–Ru Catalysts for the Rapid Hydrogenation of Toluene and Tetralin at Ambient Temperature and Pressure. *Appl. Catal. A* **2013**, *454*, 46–52. [CrossRef]
33. Song, L.; Li, X.; Wang, H.; Wu, H.; Wu, P. Ru Nanoparticles Entrapped in Mesopolymers for Efficient Liquid-Phase Hydrogenation of Unsaturated Compounds. *Catal. Lett.* **2009**, *133*, 63–69. [CrossRef]

colloids and interfaces

MDPI

Review

Biocompatible Emulsions Stabilized by Natural Silk Fibroin

Xiuying Qiao [1,*], Reinhard Miller [2], Emanuel Schneck [2] and Kang Sun [1]

1 State Key Laboratory of Metal Matrix Composites, School of Materials Science and Engineering, Shanghai Jiao Tong University, Shanghai 200240, China; ksun@sjtu.edu.cn
2 Institute for Condensed Matter Physics, Technical University of Darmstadt, 64289 Darmstadt, Germany; reinhard.miller@pkm.tu-darmstadt.de (R.M.); emanuel.schneck@pkm.tu-darmstadt.de (E.S.)
* Correspondence: xyqiao@sjtu.edu.cn

Abstract

Due to its amphiphilicity, the natural fibrous structural protein, silk fibroin (SF), can adsorb at the oil/water interface, form protective viscoelastic layers, and stabilize emulsions. Biocompatible SF-stabilized emulsions can be used in different fields of cosmetics, food, drug delivery, and biomedicine. Depending on the silk processing method, various emulsion types can be obtained, such as film-stabilized emulsions stabilized by SF molecules and Pickering emulsions stabilized by nanostructured SF or SF particles. Nanostructured SF and SF particles, with β-sheet dominated secondary structures, can overcome the drawback of SF molecules with unstable conformation transition during application, and thus endow higher emulsion stability than SF molecules. The emulsions stabilized by SF nanoparticles can endure heat and high ionic strength, while the emulsions stabilized by SF nanofibers show superior stability at high temperature, high salinity, and low pH due to the strong interfacial entangled nanofiber networks. In this review, the recent progress in research on SF-stabilized emulsions is summarized and generalized, including a systematic comparison of the stabilization mechanisms for different SF morphologies, and the influences of the emulsion fabrication technique, component type and proportions, and environmental conditions on the microstructures and properties of SF-stabilized emulsions. Understanding the stabilization mechanism and factors influencing the emulsion stability is of great significance for the design, preparation and application of SF-stabilized emulsions.

Keywords: silk fibroin; oil/water interface; emulsion; nanofibers; nanoparticles

1. Introduction

Emulsions, which help to incorporate, encapsulate, or deliver lipophilic bioactive compounds into various products, are widely utilized in the fields of food, pharmaceuticals, personal care, and cosmetics [1]. In fact, emulsions are thermodynamically unstable systems consisting of at least two immiscible liquids, one of which is dispersed in the form of droplets in the other, and they tend to break down due to the effects of gravitational separation or creaming, flocculation, drop coalescence, and Ostwald ripening [2]. However, emulsions can be stabilized by amphiphilic agents, such as traditional surfactants, proteins, or colloidal solid particles, which can reduce interfacial tension, form interfacial films, and prevent droplets from coalescing [3,4]. As a kind of novel emulsifier, proteins have attracted great attention due to their biocompatibility, biodegradability, and intrinsic amphiphilic properties [5]. It is widely accepted that proteins, which are nontoxic, abundant, and biologically sustainable, can replace synthetic surfactants to produce environmentally friendly emulsion products [6].

Generally, proteins' intrinsic properties and their concentration, as well as their environmental conditions like pH, temperature, and additives, significantly affect their surface activity, interfacial adsorbed layers, and the resulting emulsion stability [7,8]. Increasing protein concentration helps, for example, to increase the adsorption rate and structural rearrangements of bovine serum albumin (BSA) and to shorten the gelation time of monoclonal antibodies (mAb) at oil/water interfaces [9,10]. At the isoelectric pH value (pI), BSA and human immunoglobulin G exhibit a minimum interfacial elastic modulus and tension [11], and emulsions stabilized by BSA or whey proteins exhibit a larger droplet size than at other pH values due to the weakening of the electrostatic repulsions between droplets [12,13]. The adsorption layers of potato protein isolate show a major elastic response and a higher ability to produce stable emulsions at pH 8 rather than at pH 2 [14]. As for BSA, the increase in ionic strength leads to an increased interfacial elasticity, a decrease in the interfacial tension, and an easily packed compact coiled conformation due to the counter-ion screening effect [11]. β-Lactoglobulin (BLG) can form complexes with cationic alkyltrimethylammonium bromide (ATAB) surfactants and anionic sodium dodecyl sulphate (SDS) surfactant due to electrostatic interactions, but with the increase in surfactant concentration, the surfactant domains begin to govern the interfacial properties after a gradual replacement of the protein molecules from the interface [15]. Owing to the competitive adsorption of nonionic surfactants, the incorporation of Tween 20 results in a remarkable decrease in the surface shear viscosity of BLG layers at the n-tetradecane/water interface [16], and the addition of Tween 60 causes a notable reduction in the layer thickness of casein at the soy-oil/water interface [17].

Silk fibroin (SF), as a fibrous structural protein isolated from domesticated silkworms, exhibits excellent mechanical properties and biocompatibility and minimal inflammatory reactions, and therefore, it has attracted great attention in biotechnological and biomedical applications, and also in the production of functional food [18–20]. Depending on the silk processing methods, SF exists in different forms like random coil, α-helix, and crystalline β-sheets [21], and can be produced as films, gels, membranes, powders, and porous sponges [22,23]. SF consists of highly repetitive amino acid sequences with alternating hydrophobic and hydrophilic blocks along the molecular chain [24,25], and the resulting amphiphilicity, and hence surface activity, endows SF with the ability to adsorb at oil/water interfaces just like surfactant molecules, form protective viscoelastic layers, and stabilize emulsions [26–31]. Biocompatible emulsions stabilized by SF have a promising future for applications in different fields of cosmetics, food, drug delivery and biomedicine [5]. Moreover, taking advantage of the emulsions stabilized by SF, biodegradable microparticles and microspheres for drug delivery [32,33], electrospun nanofibers, and scaffolds for tissue engineering [34,35], high-strength shape-memory organohydrogels can be produced [36]. Different morphologies of SF can produce various emulsions, those stabilized by molecular SF films and Pickering emulsions stabilized by nanostructured SF and SF particles [31]. Pre-structured SF, such as SF nanofibers, SF nanobrushes, and SF micro- or nanoparticles, can overcome the drawback of SF molecules with unstable conformation transitions during application, and endow the emulsions with a higher stability. In this review, the recent progress in research on SF-stabilized emulsions is summarized and generalized, including a systematic comparison of the stabilization mechanism for SF with different morphologies, and the influences of the emulsion fabrication technique, SF concentration, oil polarity and ratio, pH, and ionic strength on the microstructures and properties of SF-stabilized emulsions. Understanding the stabilization mechanism and influence factors for the emulsion stability is of great significance for the design, preparation, and application of SF-stabilized emulsions.

https://doi.org/10.3390/colloids10010013

2. Emulsions Stabilized by Silk Fibroin Molecules

After the extraction of sericin by degumming with a Na_2CO_3 aqueous solution, the dissolution of cocoon fibers with a LiBr aqueous solution, and the dialysis against deionized water, SF aqueous solutions with a concentration of 19–21 mg/mL can be achieved from fresh domestic *Bombyx mori* cocoons [26,27,36–38]. The molecular weight of SF is 8–30 kDa, where the low-molecular-weight components result from molecular breakage during the degumming process [37]. Homogenization is a simple, high-energy method to produce emulsions [37]. By utilizing high-shear dispersion emulsifiers, SF molecules initially solved in water can emulsify different oil phases, like other proteins. With the increase in homogenization speed and time, the size of emulsion droplets decreases (see Figure 1), and the droplet coalescence slows down [37,38]. Moreover, the emulsions fabricated at higher homogenization speeds exhibit fewer irregular droplets, a faster liquid-gel transition, and less creaming during storage than those obtained at lower homogenization speeds. However, the emulsions fabricated at longer homogenization times show no obvious changes with regard to the creaming stability [37]. Thus, speeding up and prolonging the homogenization step play an important role in slowing down the droplet coalescence, and therefore are beneficial for producing more stable emulsions with smaller and more regular droplets. However, when the homogenization speed and time exceed a critical value, the emulsions change from a liquid state to a semi-gel [37].

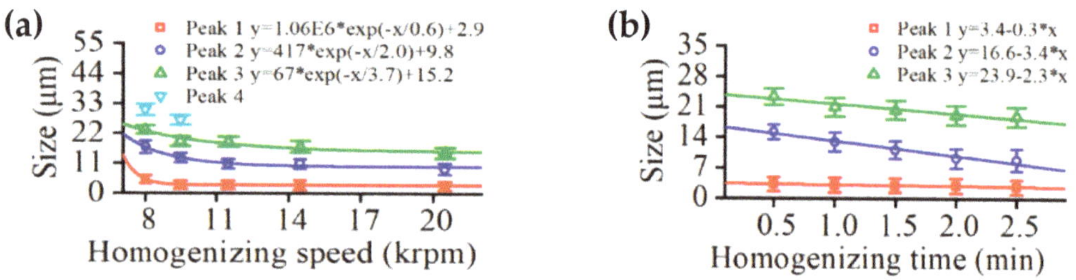

Figure 1. Droplet size of SF-stabilized fish oil/water emulsions with an oil/water volume ratio of 1:1 fabricated at different homogenizing speeds (**a**) and times (**b**). They were summarized from the Gaussian fitting of the droplet size distribution in the emulsions, and "Peak" refers to the peaks in the fitted multimodal distributions [37].

Compared to traditional surfactants, SF molecules diffuse and adsorb more slowly to the oil/water interface. The interface allows only a limited amount of SF molecules to adsorb, and further increase in the SF concentration (C_{SF}) leads to an excess of SF in the bulk or structural defects in the interfacial networks [29]. The formation of interfacial films with β-sheet secondary structures after adsorption is very slow, too, and takes several hours to attain equilibrium, especially for high C_{SF} and more polar oil phases [29]. Like lysozyme [39], not all the droplets of SF-stabilized emulsions exhibit spherical shapes, but rather are elliptical or have even more intricate shapes because of the high interfacial storage moduli (see Figure 2) [27,28,37]. In addition, an increased oil content can also result in a slight deformation of emulsion droplets at the edge due to the dense packing and squeezing of the droplets [35]. The size of emulsion droplets decreases with the increase in C_{SF} [26,36–38], which can be attributed to the enhanced protein adsorption, the strengthening of interfacial elasticity, and the resulting suppression of flow-induced coalescence, in line with the results reported for sweet potato proteins and concentrated flaxseed proteins [40,41]. The increase in droplet size with increasing oil/water volume ratio (ϕ_o) [26,35–38] is related to the increase in the total droplet surface area and the droplet collision frequency, and the decrease in the amount of adsorbed SF molecules.

https://doi.org/10.3390/colloids10010013

Taken together, all this results in less robust and protective interfacial layers and, in turn, promotes droplet coalescence [26]. When the oil content exceeds a certain percentage, the emulsification fails with oil leakage due to the insufficient loading of the droplets by protein molecules [38]. Moreover, the droplet size decreases with the increase in oil polarity from dodecane to hexanol, which can be attributed to lower corresponding interfacial tension [42]. However, the dodecane/water emulsion with larger droplets exhibits better dispersed oil droplets in the continuous water phase with less flocs, which was interpreted as a consequence of the higher interfacial modulus of the SF layers formed at the dodecane/water interface than those at other polar oil/water interfaces [26]. The decrease in initial emulsion droplet size helps to decrease emulsion creaming and improve the emulsion stability [37]. With an increased storage time, the emulsion droplets become gradually irregular and increase in size because of droplet coalescence [37].

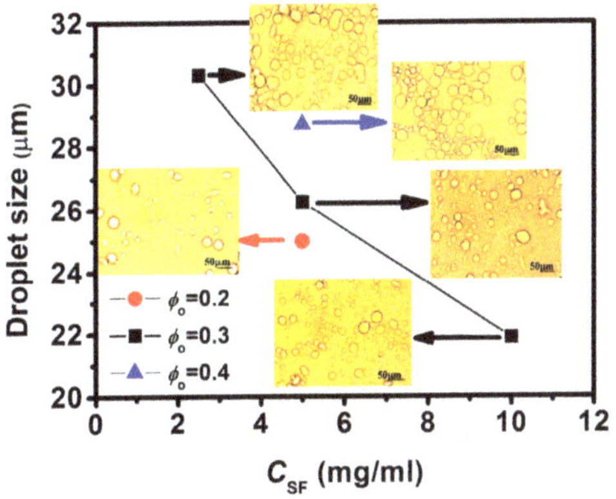

Figure 2. Dependence of droplet size on SF concentration (C_{SF}) and oil/water volume ratio (ϕ_o) for butyl butyrate/water emulsions stabilized by SF. The inset plots are microscope images of these emulsions taken 7 days after emulsion preparation [26].

In general, the emulsion stability can be evaluated by the creaming index (CI), which is obtained by dividing the serum layer height by the total emulsion height [27]. Creaming, followed by phase separation with released water and/or oil, is one of the destabilization mechanisms of emulsions caused by gravity [26]. During emulsification, there is a dynamic equilibrium between the droplet disruption and droplet coalescence. Therefore, before reaching a steady state, all O/W emulsions stabilized by SF experience an initial increase in the CI value. CI decreases with the increase in C_{SF} and ϕ_o (see Figure 3), and it is obvious that the emulsion stability is greatly improved by increasing C_{SF} and ϕ_o [26,36,37]. Before a saturated interfacial adsorption is reached, high C_{SF} helps to shorten the time to attain adsorption equilibration, and to form denser adsorption layers with stronger colloidal networks, thus leading to a higher interfacial elastic modulus, slower droplet coalescence, and less creaming [26,37,38]. Moreover, denser adsorption layers hinder the droplet movement, which results in higher elastic moduli, a complex bulk viscosity, and yield stress of the emulsions, which all together further enhance the emulsion stability [26,27]. However, as C_{SF} exceeds a critical value, the adsorbed SF molecules can no longer pack well due to the jamming state and nonideal molecular arrangements, which causes the decrease in interfacial elasticity and toughness [29]. This leads to the change in the emulsions from a liquid state to a semi-gel [37]. Higher ϕ_o leads to a closer packing and stronger interactions between the droplets, thus leading to a faster liquid-gel transition, enhanced elastic modulus, complex viscosity, and yield stress of the emulsion, and to a higher emulsion stability with less creaming [26,40,43]. However, larger ϕ_o brings about more irregular

https://doi.org/10.3390/colloids10010013

droplet shapes and faster droplet coalescence during storage, owing to the decreased SF adsorption and increased number of droplet collisions [26,37]. The high internal phase emulsions with more oil are easier to form emulsion gels and have a semi-solid texture. Emulsion semi-gels and gels are widely used in the food industry [37]. Evidently, it is the enhancement of SF adsorption and the increased packing of oil droplets that promote the modulus, yield stress, and then the emulsion stability [26,36,44]. Moreover, as for the SF-stabilized emulsions, higher oil polarity results in larger storage moduli, enhanced yield stresses, slower equilibration of emulsions, and higher emulsion stability with less water released before reaching a steady state, although the interfacial modulus of SF is lower at polar oil/water than that at nonpolar oil/water interfaces [26]. It seems that the emulsion system behavior is more complex than the single oil/water interfaces, and sometimes, the impact of SF on the interfacial viscoelasticity and emulsion stability is slightly different upon the change in preparation conditions. The butyl butyrate oil leads to a two-stage creaming process because of its spatial restriction of the ester groups, which affects interfacial molecular arrangements in ways that are negative for emulsion stability [27]. Although some water or serum is released after emulsification, the remaining emulsions are very stable over several months, owing to the protection by the adsorbed interfacial SF layers that suppress droplet coalescence. The interfacial shear rheological behavior of SF layers shows that the adsorbed SF molecules form elastic networks with β-sheet dominated structures at oil/water interfaces through conformation transitions, structural reorganization, and physical crosslinking [29,35], so that the equilibrium shear storage modulus ($G_e{}'$) of the adsorbed SF layers is much higher than those, for example, of mAb, β-casein, and BLG layers at oil/water interfaces [10,45,46]. The bulk rheological behavior of SF-stabilized emulsions indicates that well-developed three-dimensional elastic network structures also exist in emulsions [26], and the network formation should be attributed to the close packing and association of droplets [47], which can be clearly observed in the insets of Figure 2, especially for emulsions with larger C_{SF} and ϕ_o [26]. The formed strong elastic networks by SF self-assembly at the oil/water interface significantly enhanced the emulsion stability.

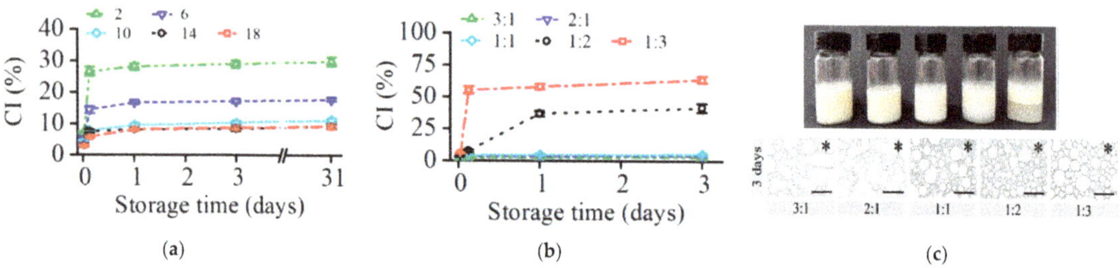

Figure 3. Change in the creaming index (CI) with storage time and images for SF-stabilized fish/oil emulsions: (**a**) CI of emulsions with the same ϕ_o (1:1) but different C_{SF} in aqueous solution (2, 6, 10, 14, and 18 mg/mL); (**b**) CI of emulsions with the same C_{SF} = 14 mg/mL in aqueous solution but different ϕ_o (3:1, 2:1, 1:1, 1:2, and 1:3); and (**c**) macroscopic and microscopic images of emulsions taken 3 days after preparation, and the scale bar is 20 μm [37].

The environmental conditions show a great influence on the stability of emulsions stabilized by SF. The stable emulsion fraction is higher for systems with pure water than with buffer, especially at pH 4, although the interfacial tension at pH 4 shows a faster and more remarkable decrease and the lowest equilibrium values. At pH 4 around the isoelectric point (pI) of SF, the charge-neutralized SF molecules no longer repel each other electrostatically and therefore aggregate more easily and form more rigid interfacial layers with compact structures [27], which result in a faster molecular adsorption at the interface and a higher modulus of interfacial layers at low C_{SF} (see Figure 4). However, the

https://doi.org/10.3390/colloids10010013

weakening of interfacial ductility directly causes an easier structural rupture and a subsequent droplet coalescence upon small emulsion perturbations [48,49], thus providing emulsions with a lower emulsion stability and larger droplet size [27]. When pH ≥ 8, the increasing electrostatic repulsion among SF molecules disrupts the adsorption of SF at oil/water interfaces, thus making the resulting emulsion less stable and even fail to form at pH 10 [38]. The emulsion droplet size shows a maximum at pH 4, close to the pI, where the absence of electrostatic repulsion between the droplets promotes coalescence [27,38]. On the other hand, the positively charged ions significantly affect the adsorption and layer properties of SF at the oil/water interface and thus the emulsion stability, as a result of the strong electrostatic interactions between counter-ions and the negatively charged groups of SF (see Figure 5) [28]. Higher ionic strength leads to lower interface tension, faster establishment of the adsorption equilibrium, higher interfacial strength, and lower interfacial fracture at large deformation, due to the reduction in the effective charge on SF molecules and enhanced adsorption and intermolecular interaction. However, at higher ion concentrations, the electrostatic screening and ion-binding effects result in SF aggregation, droplet coalescence, and a decreased emulsion stability, owing to the reduction in electrostatic repulsion. The droplet size exhibits a drastic increase with the addition of salt, and the stable emulsion fraction decreases with the increase in salt concentration (C_{ion}), especially for dodecane/water emulsions. The emulsions are more stable when the added salt is monovalent (NaCl) than when it is divalent ($CaCl_2$) or even trivalent ($NdCl_3$). This is in agreement with earlier reports on α-lactalbumin and BLG, where destabilization is attributed to Ca^{2+} binding with the free carboxylic groups of aspartic and glutamic acids. This phenomenon can be explained by the stronger elastic modulus of the interfacial SF layers in the presence of NaCl, as evaluated by interfacial shear rheology [50]. It is found that a suitable addition of negatively charged beet pectin can improve the stability of SF emulsions at pH 4 by covering the surface of positively charged SF-coated oil droplets and thus preventing emulsion aggregation. A low concentration of beet pectin causes charge neutralization and bridging flocculation, and a high concentration of beet pectin causes depletion flocculation of larger droplets, both of which lead to a sharp decline of emulsion stability. Beet pectin can also improve the stability of SF-stabilized emulsions at high salt concentrations at pH 3 by increasing the steric repulsion, that was interpreted by the authors as the consequence of a reduction in van der Waals interactions between the droplets. Besides emulsion stability, beet pectin enhances the oxidative stability of SF-stabilized emulsions during storage [49]. In practical applications, strong emulsion stability against environmental stress is generally required, because of the coexistance of emulsifiers and additives, and the change in pH and temperature. The sensitivity of SF molecule-stabilized emulsions to pH and ionic strength due to the unstable interfacial conformation changes significantly limits the applications of SF as an emulsifier. Therefore, pre-structured SF may solve this problem, endow the emulsion with a higher stability to resist environmental changes, and widen the applications of SF-stabilized emulsions in various fields.

https://doi.org/10.3390/colloids10010013

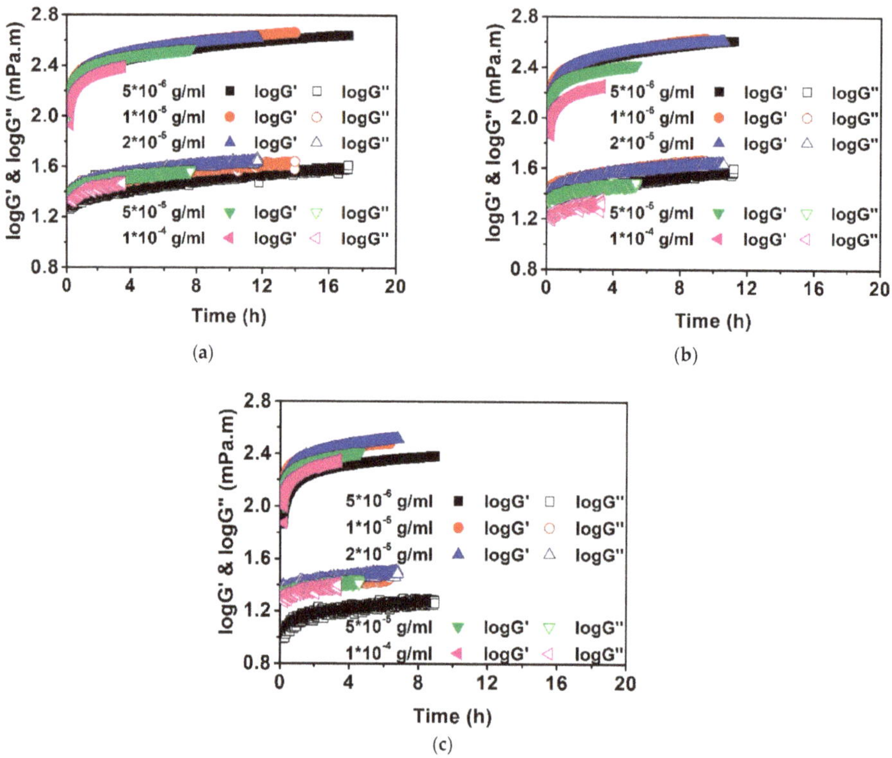

Figure 4. Time evolution of the interfacial shear moduli of SF in buffer solutions with different pH values during the adsorption at the butyl butyrate/water interface: (**a**) pH 3; (**b**) pH 4; and (**c**) pH 7. The SF concentration C_{SF} rangs from 5*10^{-6} g/ml to 1*10^{-4} g/ml [27].

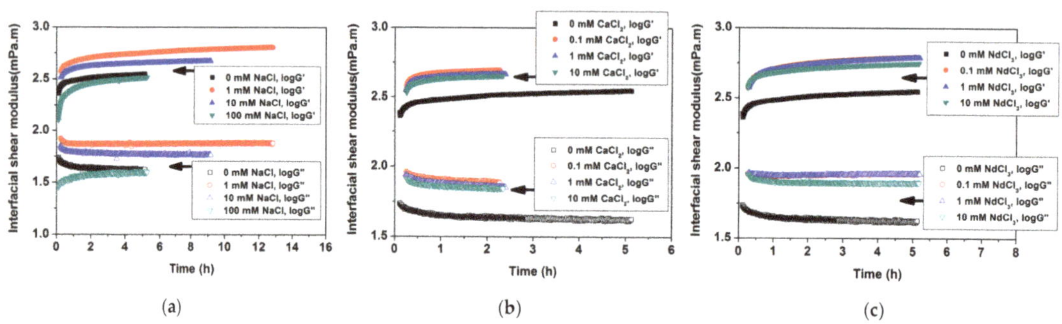

Figure 5. Time evolution of interfacial shear moduli for SF at the dodecane/water interface during adsorption from aqueous salt solutions: (**a**) NaCl; (**b**) $CaCl_2$; and (**c**) $NdCl_3$ [28].

3. Emulsions Stabilized by Silk Fibroin Nanofibers

SF nanofibers can be prepared by concentrating fresh aqueous SF solution to obtain metastable nanoparticles, diluting the concentrated solution with ultrapure water, and then incubating the diluted solution at 60 °C to induce the nanofiber formation [5]. Different from SF molecules in aqueous solution, SF nanofibers, formed through a controllable self-assembly process, have a high β-sheet content and charge density, hydrophobic properties, and water dispersibility [5]. SF nanofibers can form solutions and gels in aqueous solutions at different concentrations, and they become a versatile emulsifier presenting

https://doi.org/10.3390/colloids10010013

excellent long-term stability at high temperature, near the isoelectric point, and at high salinity [5]. Emulsions can be obtained by using high-speed dispersion machines to emulsify SF nanofiber solutions with oils [5]. Similarly to SF molecules, SF nanofibers can also form interconnected networks with pore sizes of 3–8 nm at oil/water interfaces and stabilize emulsions in a wide range of water/oil ratios and oil polarities (see Figure 6). The size of emulsion droplets first increases with the water/oil ratio but then decreases again, exhibiting a maximum at a ratio of 3:7. Larger SF nanofiber concentration, higher homogenization speed, and longer homogenization time help to decrease the droplet size. Emulsions stabilized by SF nanofibers show superior stability, being stable without any phase separation for 6 months at room temperature or even at 60 °C. Moreover, these emulsions are still stable after 3 months at high salt concentrations and low pH values due to the strong physical entanglements and interactions of SF nanofibers, a situation that surfactants, peptides, and SF molecules cannot achieve. The superior stability endows SF nanofibers with the universality and versatility for emulsion applications. Hence, tunable microcapsules can be fabricated by regulating the oil phases, the water/oil ratio, and the concentration, conformation, and size of SF nanofibers.

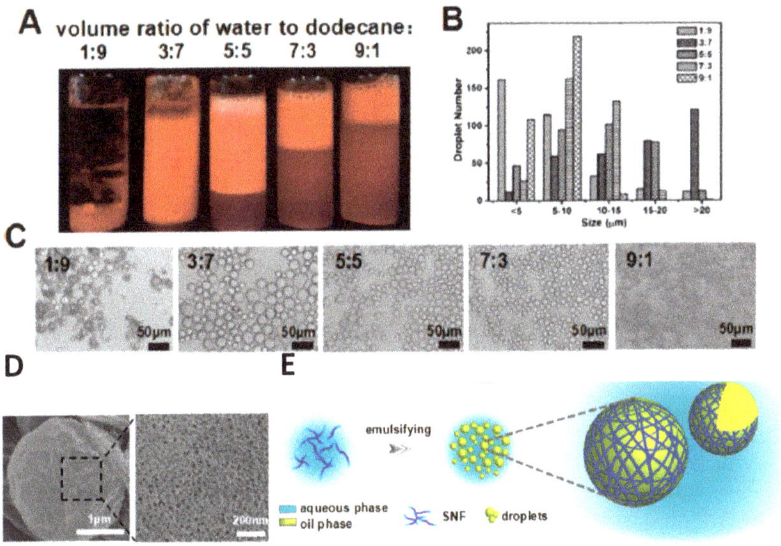

Figure 6. Morphology and structure of emulsions formed at various volume ratios of SF nanofiber solution to dodecane: (**A**) optical photographs; (**B**) size distribution; (**C**) fluorescence micrographs of the emulsions at various water/oil ratios; (**D**) high magnification of the interface in the emulsions; and (**E**) schematic of the process for the formation of oil-in-water emulsions [5].

4. Emulsions Stabilized by Silk Fibroin Nanobrushes

As a nanocomposite of SF nanofibers and SF nanowhiskers, SF nanobrushes have a shorter length than SF nanofibers and a partially thinner diameter than SF nanowhiskers, and possess peculiar three-dimensional nanostructure, excellent structure-performance-regulated and function-enhanced properties, and biocompatibility [51]. SF nanobrushes can be prepared by a nanowhisker nanotemplate-guided self-assembly of an aqueous SF solution. According to the classical micelle model of silk assembly, the SF molecules in aqueous solution self-assemble directly on the straight segmeents of SF nanowhiskers, which are isolated from degummed SF by the top-down acid-hydrolysis approach. SF nanobrushes exhibit a greater ability to lower the interfacial tension than SF nanofibers

https://doi.org/10.3390/colloids10010013

and nanowhiskers. Similarly to SF nanofibers and nanowhiskers, the β-strands of SF nanobrushes orient parallel to the fibrillar axis. The peculiar brush-like nanostructure endows SF nanobrushes with the capability to form much more sophisticated and interconnected networks than straight rod/fiber-like nanomaterials when stabilizing oil/water interfaces, and ultra-stable biocompatible emulsions can be achieved by ultrasonicating an aqueous SF nanobrush dispersion and an oil phase. SF nanobrushes can resist creaming in a wide range of concentrations and stabilize various types of emlsions from liquid-like to gel-like ones. Higher ultrasonication intensity leads to the breakage of more fragments from the branches at SF nanobrushes and larger droplet size (see Figure 7), thus causing the decrease in the emulsion stability. Increasing SF nanobrush concentration helps to reduce the creaming index, decrease the droplet size, and enhance the emulsion viscosity, due to the strong interactions and interchain network formation of the relatively long and flexible SF nanofiber branches in the nanobrushes. However, when the SF nanobrush concentration exceeds a critical value, droplet flocculation occurs due to the formation of "strips" by the linkage of excess SF nanobrushes. Moreover, emulsions stabilized by SF nanobrushes exhibit gel-like and shear-thinning behaviors in rheological measurement [51].

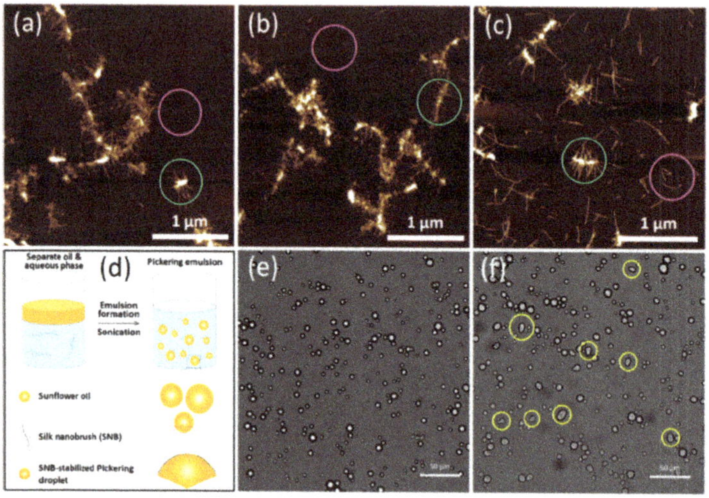

Figure 7. AFM images of SF nanobrushes that experienced 0 (**a**), 40% (**b**), and 60% (**c**) intensity ultrasonication; schematic representation of a SF-nanobrush-stabilized Pickering emulsion preparation (**d**); SF nanobrush-stabilized Pickering emulsions formulated by 40% (**e**) and 60% (**f**) intensity ultrasonication [51]. The 0%, 40%, and 60% intensity ultrosonication likely refers to the percentage of the maximum intensity achieved by the instrument. The green circles in (**a**–**c**) show the brush-like morphology of SF nanobrushes. The pink circles in (**a**–**c**) show the incomplete brush-like morphology or even non-brushlike shape of SF nanobrushes after a large number of fragments of branches were exfoliated. The yellow circles in (**f**) show that droplets from the 60% intensity were on the verge of break, indicating the instability of the emulsions.

5. Emulsions Stabilized by Silk Fibroin Particles

In order to avoid gelation issues at high SF concentrations during emulsion fabrication, SF micro- or nanoparticles were chosen to prepare oil/water Pickering emulsions. SF microparticles, synthesized via a salting-out process by the addition of a cold solution of concentrated ionic phosphate to a SF aqueous solution to form nuclei, the Ostwald ripening in an ice bath to increase the homogeneity of particles, and the exposure to a 60 °C water bath to increase the β-sheet content and reduce particle solubility (see Figure 8). The obtained SF microparticles have a porous inner structure with a pore diameter of 0.1–0.2 μm

https://doi.org/10.3390/colloids10010013

and a controllable size via regulating SF concentrations [52]. Emulsions stabilized by SF microparticles show a three-layer structure, with larger droplets in the upper layer, smaller droplets in the lower layer, and free SF microparticles at the bottom. Increasing the oil/water ratio leads to a decrease in the volume of the stable emulsion, a linear increase in the droplet size, and a higher polydispersity of droplets [52].

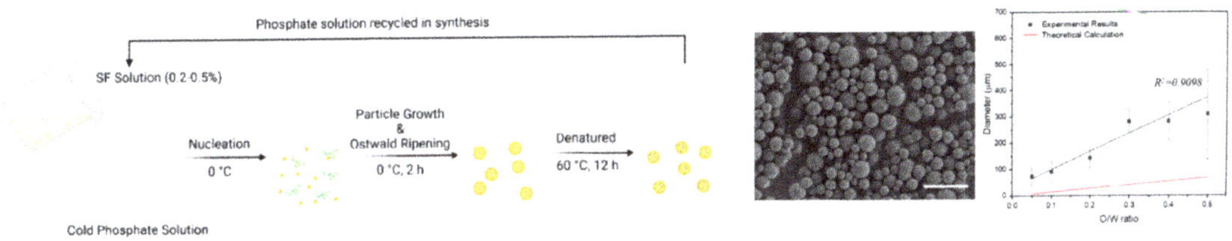

(a) (b) (c)

Figure 8. (**a**) Scheme of the synthesis procedure of water-insoluble SF microparticles. (**b**) Scanning electron microscope (SEM) image of SF microparticles synthesized in freshly prepared phosphate salt solutions, showing a diameter range of 2–3 μm. (**c**) Linear fitting of the Pickering emulsion droplet diameter increasing with the oil/water ratio [52].

SF nanoparticles, with a folded β-sheet-dominated secondary structure, can be prepared by dissolving degummed silk in non-toxic, non-volatile, and cost-effective aqueous phosphoric acid (PA), and then by coagulating in acetone [38]. The β-sheet structures help SF nanoparticles to enhance the structure stability against environmental stresses such as heating and high salt concentrations, and to prevent emulsion destabilization caused by the conformational changes in SF. The three-phase contact angle of 92.5°, very close to 90°, renders SF nanoparticles highly capable to adsorb at oil/water interfaces. The negtive zeta potential of around −20 mV endows SF nanoparticles with the ability to prevent droplet coalescence and to enhance the emulsion stability by electrostatic repulsion (see Figure 9). During emulsion fabrication, increasing homogenization pressure and time results in a stronger emulsification capability of SF nanoparticles and finer emulsions with smaller droplet size. After experiencing creaming, the obtained Pickering emulsions are very stable with few changes in droplet size after storing for one month. The size of emulsion droplets decreases with the increase in particle load, homogenization time, homogenization speed, and the decrease in oil/water ratio. More stable emulsions can form with smaller SF nanoparticles, higher nanoparticle loading, and lower oil volume fractions. Similarly to the emulsions stabilized by SF molecules, the Pickering emulsions stabilized by SF nanoparticles show gel-like structure characteristics, larger storage modulus and viscosity at higher emusifier loading, and a shear-thinning behavior. Moreover, the Pickering emulsions stabilized by SF nanoparticles can endure heat and high ionic strength, exhibiting no significant change in droplet size after storage at 60 °C and adding 1 M salt for one week. However, the stability of these emulsions is very sensitive to pH due to the changes in the surface charge of SF nanoparticles. At pH 2 and pH 6, the emulsions can be stabilized by repulsive interactions between the positively charged or negtively charged particles, respectively. At pH 4, near the isoelectric point of SF, the charge neutralization makes SF nanoparticles lose mutual repulsion, which in turn leads to the increase in droplet size and a decreased emulsion stability. At alkaline pH (pH ≥ 8), the extremely strong electrostatic repulsion among the particles disrupts the interfacial particle adsorption, and even results in particle desorption from the interface and final emulsion breakdown at pH 10 [38]. The pH-responsiveness makes these Pickering emulsions useful for emulsion polymerization and catalyst/emulsifier recycling [53].

https://doi.org/10.3390/colloids10010013

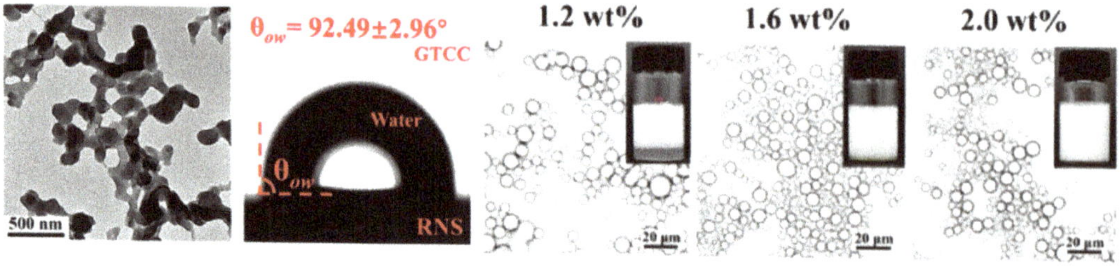

(a) (b) (c)

Figure 9. (**a**) TEM image and (**b**) three-phase contact angle of SF nanoparticles. (**c**) The microstructures and bulk appearances of SF nanoparticles-stabilized emulsions with the same oil/water ratio (1:9) but different particle concentrations 1 day after preparation [38].

6. Stabilization Mechanism of SF-Stabilized Emulsions

Like other proteins, SF molecules can stabilize emulsions by adsorbing at the oil/water interface, decreasing the interfacial tension, forming protective films at the droplet surface, and preventing or decelerating droplet coalescence and/or Ostwald ripening. In addition, the SF network structures formed at the interface increase the emulsion viscosity and reduce the encountering frequency of droplets, further contributing to the stabilization of emulsions. At the oil/water interface, the adsorbed SF molecules experience conformational transitions and intermolecular self-assembly and thereby form stable, solid-like interfacial gels with ordered β-sheet structures and high interfacial moduli [35]. The conformational changes and molecular rearrangements of SF seem very fast without obvious interfacial gel transitions, but the β-sheet formation experiences much slower physical crosslinking and structural reorganization into entangled networks. Higher SF concentrations help to shorten the time to attain adsorption equilibration and to form network structures at the interface with a higher elastic modulus that resist breakage, which leads to a higher emulsion stability with less water released during storage. However, as C_{SF} exceeds a critical value, the adsorbed SF molecules can no longer pack well due to the jamming state, thus causing the decrease in interfacial elasticity and toughness and even the transition of the emulsion from a liquid to gel state. Interfacial rheology results show that the interface between water and a nonpolar oil, in comparison to a polar oil, supports more SF molecules and promotes the formation of interfacial SF films with higher elastic moduli and greater toughness [29]. However, macroscopic emulsions with high-polarity oil exhibit higher stability, larger storage moduli, and enhanced yield stresses in the bulk rheology [26]. This indicates that the interfacial layers formed in bulk oil/water systems may be slightly different from the interfacial layers formed at single oil/water interfaces, bringing about the differences in the impact of oil polarity on the microscopic interfacial stability and the macroscopic emulsion stability.

The stabilization of emulsions with molecular SF as an emulsifier depends on the interfacial self-assembly of SF molecules to form β-sheet structures, while SF nanofibers with typical β-sheet structures can be directly used to stabilize emulsions [5]. SF nanofibers can form tight physical entanglement networks at the oil/water interface, and endow superior stability. The conformation of SF nanofibers shows no change after emulsion formation, which overcomes the drawback of single SF molecules with uncontrollable conformational transition. Hydrophobic SF nanofibers provide a controllable and reliable material to prepare more stable emulsions, as compared not only with single SF molecules, but also with other peptide nanofibers. The higher stability of emulsions is reached by SF nanofibers with respect to SF molecules because of their tighter and stronger interfacial networks and the force balance between a charge repulsion inside and outside the droplets.

https://doi.org/10.3390/colloids10010013

At neutral state and without salt addition, the decreased charge repulsion outside the droplets weakens the interaction balance and emulsion stability. With the addition of salt or acid, the charge repulsion inside and outside the droplets is shielded, and the dominant physical entanglements and interactions of SF nanofibers help to recover the emulsion stability. The pore size of the interfacial SF nanofiber networks decreases at higher salt concentrations and lower pH values, and the reduction in the charge repulsion among SF nanofibers strengthens nanofiber interactions and interfacial network structures [5]. The superior stability of SF nanofiber-stabilized emulsions at high temperature, high salt concentrations, and low pH values helps to widen their applications in various fields. SF nanobrushes, as a nanocomposite of SF nanowhisker backbones and SF nanofiber branches, have a lower interfacial tension than the SF nanofibers and SF nanowhiskers alone. Their peculiar three-dimensional brush-like nanostructure allows SF nanobrushes to form much more sophisticated and interconnected networks than SF molecules and simple SF nanofibers at the oil/water interface, and thus ultra-stable biocompatible emulsions can be obtained for a wide range of concentrations [51].

Different from SF molecules and SF nanofibers, SF particles stabilize emulsions by forming dense particle layers after irreversible adsorption at the oil/water interface. The resulting strong mechanical barrier prevents droplet flocculation and coalescence. Moreover, the surface charge of SF particles, contributed by the amino-acid sequences of the hydrophilic domains, generates electrostatic repulsion between the emulsion droplets, thus preventing droplet aggregation and enhancing emulsion stability [38]. SF particles obtain the folded β-sheet secondary structure through the self-assembly of SF molecules during the regeneration process, similar to SF nanofibers and SF nanobrushes. The water-insoluble β-sheets, formed by the folding of hexapeptide sequences in SF molecules, enhance the structural stability against environmental stress and prevent emulsion destabilization [38]. Therefore, the pre-structuring procedure avoids the unstable structure transition of SF molecules in actual applications, and the emulsions stabilized by SF nanofibers and SF nanoparticles show superior stability at high temperature and high salt concentration.

7. Conclusions

SF concentration, SF morphology, oil polarity, oil/water volume ratio, and environmental conditions play significant roles in the molecular adsorption, structural reorganization, and interfacial viscoelasticity of SF at oil/water interfaces. In turn, these factors, as well as the emulsion fabrication technique, have great influences on the stability of the resulting emulsions. Increasing homogenization speed, time, or pressure helps to promote the emulsification efficiency and to produce more stable emulsions with smaller droplet size and fewer irregular droplets. The time to attain the adsorption equilibrium is shortened with increasing SF concentration, with increasing oil polarity, and with the addition of salt. The interfacial strength and toughness of SF layers and, hence, the emulsion stability can be enhanced by raising SF concentration and adding salt to the aqueous phase. pH values close to the pI or above pH 8 lead to emulsion instability, due to the loss of mutual repulsion caused by charge neutralization or SF desorption caused by the extremely strong electrostatic repulsion.

Nanostructured SF, such as SF nanofibers and SF nanoparticles, possesses β-sheet-dominated secondary structures, and the enhanced structural stability endows them with the capability to resist environmental stresses such as high temperature and high salt concentrations, which overcomes the drawback of isolated SF molecules with unstable conformation transitions during the application. Emulsions stabilized by SF nanofibers can even show superior stability at low pH, because of the formation of tight and strong entangled networks, which endows SF nanofibers with the universality and versatility for

https://doi.org/10.3390/colloids10010013

various emulsion applications. Although the emulsions stabilized by SF nanoparticles cannot resist extreme pHs, the pH-responsiveness can endow these Pickering emulsions with intelligent features that are very useful for emulsion polymerization and catalyst/emulsifier recycling. SF molecules stabilize emulsions by interfacial molecular self-assembly to β-sheet-structured viscoelastic films, after adsorbing at oil/water interfaces. Nanostructured SF and SF particles, with β-sheet dominated secondary structures, can be directly used to stabilize emulsions by physical entanglement networks or particle adsorption layers formed at the oil/water interface. Compared with other protein-stabilized emulsions, the SF-stabilized emulsions possess great advantages of low cost, controllable microstructure and properties, higher compatibility with biocomponents, and superior stability at high temperatures, high salt concentrations, and extreme pH values, via the regulation of the SF morphology. Biocompatible emulsions stabilized by SF have a promising future for applications in different fields, such as in cosmetics, food, drug delivery, and biomedicine. Some functional food, skin care products, and porous scaffolds have already been commercialized [20,54]. The understanding of the relationship between adsorption, interfacial viscoelasticity, and the emulsification properties of SF is very important for the design, preparation, and application of SF emulsions in various fields.

Author Contributions: Conceptualization, X.Q. and R.M.; methodology, X.Q. and R.M.; investigation, X.Q.; resources, E.S. and K.S.; writing—original draft preparation, X.Q.; writing—review and editing, X.Q., R.M., and E.S.; and project administration, E.S. All authors have read and agreed to the published version of the manuscript.

Funding: This research was funded by the ALEXANDER VON HUMBOLDT FOUNDATION, grant number CHN/1150450.

Data Availability Statement: The original contributions presented in this study are included in the cited references. Further inquiries can be directed to the corresponding author.

Acknowledgments: Reproduced from Refs. [26,27,36] with permission from Elsevier. Reproduced from Ref. [28] with permission from Wiley. Reproduced from Refs. [5,34,35,50] with permission from ACS.

Conflicts of Interest: The authors declare no conflicts of interest.

References

1. Miller, R.; Guzmán, E. (Eds.) Emulsions: From single interfaces to applications. In *Progress in Colloid and Interface Science*; CRC Press: Boca Raton, FL, USA, 2025; Volume 8, ISBN 9781032636108.
2. McClements, D.J. *Food Emulsions: Principles, Practice and Techniques*; CRC Press: Boca Raton, FL, USA, 1999.
3. Dickinson, E. Food emulsions and foams: Stabilization by particles. *Curr. Opin. Colloid Interface Sci.* **2010**, *15*, 40–49.
4. Bos, M.A.; Vliet, T.V. Interfacial rheological properties of adsorbed protein layers and surfactants: A review. *Adv. Colloid Interface Sci.* **2001**, *91*, 437–471.
5. Cheng, Q.Q.; Zhang, B.B.; He, Y.; Lu, Q.; Kaplan, D.L. Silk nanofibers as robust and versatile emulsifiers. *ACS Appl. Mater. Interfaces* **2017**, *9*, 35693–35700. [CrossRef]
6. Zhang, T.; Xu, J.M.; Chen, J.H.; Wang, Z.Q.; Wang, X.C.; Zhong, J. Protein nanoparticles for Pickering emulsions: A comprehensive review on their shapes, preparation methods, and modification methods. *Trends Food Sci. Technol.* **2021**, *113*, 26–41. [CrossRef]
7. Peng, D.F.; Yang, J.C.; Li, J.; Tang, C.; Li, B. Foams stabilized by β-lactoglobulin amyloid fibrils: Effect of pH. *J. Agric. Food Chem.* **2017**, *65*, 10658–10665. [CrossRef]
8. Dombrowski, J.; Gschwendtner, M.; Saalfeld, D.; Kulozik, U. Salt-dependent interaction behavior of β-lactoglobulin molecules in relation, to their surface and foaming properties. *Colloid Surf. A Physicochem. Eng. Asp.* **2018**, *558*, 455–462. [CrossRef]
9. Tang, C.H.; Shen, L. Dynamic adsorption and dilatational properties of BSA at oil/water interface: Role of conformational flexibility. *Food Hydrocoll.* **2015**, *43*, 388–399. [CrossRef]
10. Mehta, S.B.; Lewus, R.; Bee, J.S.; Randolph, T.W.; Carpenter, J.F. Gelation of a monoclonal antibody at the silicone oil-water interface and subsequent rupture of the interfacial gel results in aggregation and particle formation. *J. Pharm. Sci.* **2015**, *104*, 1282–1290. [CrossRef] [PubMed]

11. Burgess, D.J.; Sahin, N.O. Interfacial rheological and tension properties of protein films. *J. Colloid Interface Sci.* **1997**, *189*, 74–82. [CrossRef]
12. Kulmyrzaev, A.A.; Chubert, H. Influence of KCl on the physicochemical properties of whey protein stabilized emulsions. *Food Hydrocoll.* **2004**, *18*, 13–19. [CrossRef]
13. Saito, M.; Yin, L.J.; Kobayashi, I.; Nakajima, M. Comparison of stability of bovine serum albumin-stabilized emulsions prepared by micro-channel emulsification and homogenization. *Food Hydrocoll.* **2006**, *20*, 1020–1028. [CrossRef]
14. Romero, A.; Beaumal, V.; David-Briand, E.; Cordobés, F.; Guerrero, A.; Anton, M. Interfacial and oil/water emulsions characterization of potato protein isolate. *J. Agric. Food Chem.* **2011**, *59*, 9466–9474. [CrossRef]
15. Pradines, V.; Fainerman, V.B.; Aksenenko, E.V.; Krägel, J.; Wüstneck, R.; Miller, R. Adsorption of protein-surfactant complexes at the water/oil interface. *Langmuir* **2011**, *27*, 965–971. [CrossRef]
16. Courthaudon, J.L.; Dickinson, E.; Matsumura, Y.; Clark, D.C. Competitive adsorption of β-lactoglobulin + Tween 20 at the oil-water interface. *Colloids Surf.* **1991**, *56*, 293–300.
17. Dalgleish, D.G.; Srinivasan, M.; Singh, H. Surface properties of oil-in-water emulsion droplets containing casein and Tween 60. *J. Agric. Food Chem.* **1995**, *43*, 2351–2355.
18. Pérez-Rigueiro, J.; Viney, C.; Llorca, J.; Elices, M. Silkworm silk as an engineering material. *J. Appl. Polym. Sci.* **1998**, *70*, 2439–2447. [CrossRef]
19. Lee, S.M.; Cho, D.; Park, W.H.; Lee, S.G.; Han, S.O.; Drzal, L.T. Novel silk/poly(butylene succinate) biocomposites: The effect of short fibre content on their mechanical and thermal properties. *Comp. Sci. Technol.* **2005**, *65*, 647–657.
20. Xu, L.; Wu, C.; Yap, P.L.; Losic, D.; Zhu, J.; Yang, Y.; Qiao, S.; Ma, L.; Zhang, Y.; Wang, H. Recent advances of silk fibroin materials: From molecular modification and matrix enhancement to possible encapsulation-related functional food applications. *Food Chem.* **2024**, *438*, 137964. [PubMed]
21. Sangkert, S.; Meesane, J.; Kamonmattayakul, S.; Chai, W.L. Modified silk fibroin scaffolds with collagen/decellularized pulp for bone tissue engineering in cleft palate: Morphological structures and biofunctionalities. *Mater. Sci. Eng.* **2015**, *58*, 87–88.
22. Elia, R.; Michelson, C.D.; Perera, A.L.; Brunner, T.F.; Harsono, M.; Leisk, G.G.; Kugel, G.; Kaplan, D.L. Electrodeposited silk coatings for bone implants. *J. Biomed. Mater. Res. B Appl. Biomater.* **2014**, *103*, 1602–1609. [CrossRef]
23. Zhong, T.Y.; Jiang, Z.J.; Wang, P.; Bie, S.Y.; Zhang, F.; Zuo, B.Q. Silk fibroin/copolymer composite hydrogels for the controlled and sustained release of hydrophobic/hydrophilic drugs. *Int. J. Pharm.* **2015**, *494*, 264–270. [CrossRef]
24. Jin, H.J.; Kaplan, D.L. Mechanism of silk processing in insects and spiders. *Nature* **2003**, *424*, 1057–1061. [CrossRef]
25. Vollrath, F.; Knight, D.P. Liquid crystalline spinning of spider silk. *Nature* **2001**, *410*, 541–548. [CrossRef]
26. Qiao, X.Y.; Wang, L.J.; Shao, Z.Z.; Sun, K.; Miller, R. Stability and rheological behaviors of different oil/water emulsions stabilized by natural silk fibroin. *Colloids Surf. A-Physicochem. Eng. Asp.* **2015**, *475*, 84–93. [CrossRef]
27. Tang, X.X.; Qiao, X.Y.; Sun, K. Effect of pH on the interfacial viscoelasticity and stability of the silk fibroin at the oil/water interface. *Colloids Surf. A-Physicochem. Eng. Asp.* **2015**, *486*, 86–95. [CrossRef]
28. Tang, X.X.; Qiao, X.Y.; Miller, R.; Sun, K. Effect of ionic strength on the interfacial viscoelasticity and stability of silk fibroin at the oil/water interface. *J. Sci. Food Agric.* **2016**, *96*, 4918–4928. [CrossRef]
29. Qiao, X.Y.; Miller, R.; Sun, K. Interfacial adsorption, viscoelasticity and recovery of silk fibroin layers at different oil/water interface. *Colloids Surf. A-Physicochem. Eng. Asp.* **2017**, *519*, 179–186. [CrossRef]
30. Qiao, X.Y.; Miller, R.; Schneck, E.; Sun, K. Effect of surfactants on the interfacial viscoelasticity and stability of silk fibroin at different oil–water interfaces. *J. Sci. Food Agric.* **2024**, *104*, 2928–2936. [CrossRef]
31. Mahboubi, F.N.; Simon, L.; Devoisselle, J.M.; Begu, S. From silk components to emulsions. *Adv. Colloid Interface Sci.* **2025**, *343*, 103547. [CrossRef] [PubMed]
32. Baimark, Y. Morphology and thermal stability of cross-linked silk fibroin microparticles prepared by the water-in-oil emulsion solvent diffusion method. *Asia-Pac. J. Chem. Eng.* **2012**, *7*, S112–S117. [CrossRef]
33. Cheng, C.; Teasdale, I.; Brüggemann, O. Stimuli-responsive capsules prepared from regenerated silk fibroin microspheres. *Macromol. Biosci.* **2014**, *14*, 807–816. [CrossRef]
34. Li, L.H.; Qian, Y.N.; Jiang, C.; Lv, Y.G.; Liu, W.Q.; Zhong, L.; Cai, K.Y.; Li, S.; Yang, L. The use of hyaluronan to regulate protein adsorption and cell infiltration in nanofibrous scaffolds. *Biomaterials* **2012**, *33*, 3428–3445. [CrossRef] [PubMed]
35. Wen, J.; Yao, J.; Chen, X.; Shao, Z. Silk fibroin acts as a self-emulsifier to prepare hierarchically porous silk fibroin scaffolds through emulsion−ice dual templates. *ACS Omega* **2018**, *3*, 3396−3405. [CrossRef] [PubMed]
36. Oral, C.B.; Yetiskin, R.; Cil, C.; Kok, F.N.; Okay, O. Silk fibroin-based shape-memory organohydrogels with semicrystalline microinclusions. *ACS Appl. Bio Mater.* **2023**, *6*, 1594–1603. [CrossRef] [PubMed]
37. Huang, S.; Peng, J.; Zi, Y.; Zheng, Y.; Xu, J.; Gong, H.; Kan, G.; Wang, X.; Zhong, J. Regenerated silk fibroin for the stabilization of fish oil-loaded Pickering emulsions. *Colloids Surf. A-Physicochem. Eng. Asp.* **2023**, *678*, 132523. [CrossRef]
38. Sun, S.; Deng, Y.; Sun, F.; Mao, Z.; Feng, X.; Sui, X.; Liu, F.; Zhou, X.; Wang, B. Engineering regenerated nanosilk to efficiently stabilize pickering emulsions. *Colloids Surf. A-Physicochem. Eng. Asp.* **2022**, *635*, 128065. [CrossRef]

https://doi.org/10.3390/colloids10010013

39. Erni, P.; Fischer, P.; Windhab, E.J. Role of viscoelastic interfaces in emulsion rheology and drop deformation. *J. Cent. South Univ. Technol.* **2007**, *14*, 246–249.
40. Guo, Q.; Mu, T.H. Emulsifying properties of sweet potato protein: Effect of protein concentration and oil volume fraction. *Food Hydrocoll.* **2011**, *25*, 98–106. [CrossRef]
41. Wang, B.; Li, D.; Wang, L.J.; Özkan, N. Effect of concentrated flaxseed protein on the stability and rheological properties of soybean oil-in-water emulsions. *J. Food Eng.* **2010**, *96*, 555–561. [CrossRef]
42. Wang, L.J.; Xie, H.E.; Qiao, X.Y.; Goffin, A.; Hodgkinson, T.; Yuan, X.F.; Sun, K.; Fuller, G.G. Interfacial rheology of natural silk fibroin at air/water and oil/water interfaces. *Langmuir* **2012**, *28*, 459–467. [CrossRef]
43. Torres, L.G.; Iturbe, R.; Snowden, M.J.; Chowdhry, B.Z.; Leharne, S.A. Preparation of o/w emulsions stabilized by solid particles and their characterization by oscillatory rheology. *Colloids Surf. A-Physicochem. Eng. Asp.* **2007**, *302*, 439–448. [CrossRef]
44. Dickinson, E.; Golding, M. Rheology of sodium caseinate stabilized oil-in-water emulsions. *J. Colloid Interface Sci.* **1997**, *191*, 166–176.
45. Partanen, R.; Forssell, P.; Mackie, A.; Blomberg, E. Interfacial cross-linking of β-casein changes the structure of the adsorbed layer. *Food Hydrocoll.* **2013**, *32*, 271–277. [CrossRef]
46. Jung, J.M.; Gunes, D.Z.; Mezzenga, R. Interfacial activity and interfacial shear rheology of native β-lactoglobulin monomers and their heat-induced fibers. *Langmuir* **2010**, *26*, 15366–15375. [CrossRef]
47. Erçelebi, E.; Ibanoğlu, E. Rheological properties of whey protein isolate stabilized emulsions with pectin and guar gum. *Eur. Food Res. Technol.* **2009**, *229*, 281–286. [CrossRef]
48. Tcholakova, S.; Denkov, N.D.; Ivanov, I.B.; Campbell, B. Coalescence stability of emulsions containing globular milk proteins. *Adv. Colloid Interface Sci.* **2006**, *123*, 259–293. [CrossRef] [PubMed]
49. Chen, B.; Li, H.; Ding, Y.; Rao, J. Improvement of physicochemical stabilities of emulsions containing oil droplets coated by non-globular protein-beet pectin complex membranes. *Food Res. Int.* **2011**, *44*, 1468–1475. [CrossRef]
50. Pappas, C.P.; Rothwell, J. The effects of heating, alone or in the presence of calcium or lactose, on calcium binding to milk proteins. *Food Chem.* **1991**, *42*, 183–201. [CrossRef]
51. Hu, Y.L.; Zou, Y.J.; Ma, Y.; Yu, J.; Liu, L.; Chen, M.J.; Ling, S.J.; Fan, Y.M. Formulation of silk fibroin nanobrush-stabilized biocompatible Pickering emulsions. *Langmuir* **2022**, *38*, 14302–14312. [CrossRef]
52. Hu, Y.; Cao, Y.T.; Nguyen, F.M.; Frank, B.D.; Kalinowski, M.J.; Li, M.; Rajani, S.; Marelli, B. Antibody-targeted phytohormone delivery using foliar sprayed silk fibroin Pickering emulsions. *Adv. Funct. Mater.* **2024**, *34*, 2402618.
53. Tang, J.; Quinlan, P.J.; Tam, K.C. Stimuli-responsive Pickering emulsions: Recent advances and potential applications. *Soft Matter* **2015**, *11*, 3512–3529. [CrossRef] [PubMed]
54. Li, S.J.; Chen, H.; Dan, X.; Ju, Y.K.; Li, T.; Liu, B.; Li, Y.; Lei, L.J.; Fan, X. Silk fibroin for cosmetic dermatology. *Chem. Eng. J.* **2025**, *506*, 159986. [CrossRef]

https://doi.org/10.3390/colloids10010013

colloids and interfaces

Article

Enhanced Lipid-Based Nanofungicide Formulation for Effective Control of *Ganoderma boninense* in Oil Palm

Azren Aida Asmawi [1,2], Nur Ain Izzati Mohd Zainudin [3], Nurul Aini Mohd Azman [4], Fatmawati Adam [4,5], Nurul Farhana Ahmad Aljafree [2,6], Mohamad Firdaus Ahmad [7] and Mohd Basyaruddin Abdul Rahman [2,6,*]

1 Faculty of Pharmacy and Biomedical Sciences, MAHSA University, Bandar Saujana Putra, Jenjarom 42610, Selangor, Malaysia; azrenaida@mahsa.edu.my
2 Department of Chemistry, Faculty of Science, Universiti Putra Malaysia, Serdang 43400, Selangor, Malaysia; nurulfarhana9207@gmail.com
3 Department of Biology, Faculty of Science, Universiti Putra Malaysia, Serdang 43400, Selangor, Malaysia; ainizzati@upm.edu.my
4 Faculty of Chemical and Process Engineering Technology, Universiti Malaysia Pahang Al-Sultan Abdullah, Lebuhraya Persiaran Tun Khalil Yaakob, Kuantan 26300, Pahang, Malaysia; ainiazman@umpsa.edu.my (N.A.M.A.); fatmawati@umpsa.edu.my (F.A.)
5 Centre for Research in Advanced Fluid and Processes, Universiti Malaysia Pahang Al-Sultan Abdullah, Kuantan 26300, Pahang, Malaysia
6 Foundry of Reticular Materials for Sustainability, Institute of Nanoscience and Nanotechnology, Universiti Putra Malaysia, Serdang 43400, Selangor, Malaysia
7 Department of Plant Protection, Faculty of Agriculture, Universiti Putra Malaysia, Serdang 43400, Selangor, Malaysia; firdaus1438@gmail.com
* Correspondence: basya@upm.edu.my

Abstract

Palm oil is a major agricultural commodity and an important economic driver in Asia. However, the sustainability and productivity of this crop are constantly threatened by a range of pathogenic fungi, especially *Ganoderma boninense*. Therefore, this study aimed to develop an eco-friendly hexaconazole-loaded nanoemulsion (Hexa-NE) for effective and targeted fungicide delivery while reducing environmental and health impacts. The optimized Hexa-NE formulation was evaluated for particle size, polydispersity index (PDI), zeta potential, pH, viscosity, and morphology using Transmission Electron Microscopy (TEM) and Scanning Electron Microscopy (SEM). Fungicide release, stability, and antifungal activity were conducted to assess the overall efficacy and performance of the formulation. The Hexa-NE exhibited particle size of 105.8 nm, a PDI of 0.358, a zeta potential of −53.53 mV. The formulation remained stable over three months of storage. It also demonstrated favourable physicochemical properties including low viscosity (30.24 mPa·s), low surface tension (23.87 mN/m), and suitable pH (6.14) for foliar application. TEM and SEM analyses confirmed spherical droplets and revealed significant hyphal damage to *G. boninense*. The antifungal test showed a higher inhibition of 97.1% at 0.1 μM of Hexa-NE as compared to hexaconazole solution which only 40% at the same concentration. Release studies exhibited a sustained release of hexaconazole, which may prolonged fungicidal activity. In conclusion, Hexa-NE showed promising laboratory-scale antifungal performance against *G. boninense*. These findings support its potential for further investigation as a nanoformulated fungicide for future greenhouse and field evaluations.

Keywords: nanoemulsion; nanofungicide; *Ganoderma boninense*; basal stem rot; oil palm

https://doi.org/10.3390/colloids10020024

1. Introduction

Palm oil dominates the global vegetable oil market and accounts for approximately 35.7% of total consumption. The use of palm oil was found to outweigh soybean, rapeseed, and sunflower oils [1]. Global demand for palm oil is expected to continue increasing, due to the rising need for vegetable oils worldwide [2]. Despite this increase, the sustainability and productivity of oil palm cultivation are continuosly threatened by various pathogenic fungi including *Ganoderma boninense*. This fungus is the causative agent of basal stem rot (BSR) which is considered to be the most devastating fungal disease in Southeast Asia to date [3,4]. BSR has caused a significant reduction in the density of oil palm trees per hectare and can decrease in the productivity of fresh fruit bunches up to 80% in advanced stages of infection. This results in substantial yield losses [5]. Conventional control strategies rely heavily on the use of chemical fungicides. However, their widespread use has raised environmental and health concerns [5,6]. This challenge shows the urgent need to obtain an alternative, efficient, and sustainable approaches for managing this fungal disease.

Nanotechnology has emerged as a promising platform to improve the delivery efficiency of agrochemicals. The system offers targeted delivery, enhanced penetration, and controlled release properties. Among them, nanosized lipid-based emulsions have gained increasing attention as colloidal delivery systems due to their favorable biocompatibility, ease of preparation and ability to solubilize hydrophobic active compounds. The small droplet size of lipid-based emulsions provides a high surface-area-to-volume ratio. This characteristic enhances wetting, spreading, and adhesion on plant surfaces, which in turn facilitates more efficient uptake and prolonged retention of the active ingredient [7,8]. Besides, Campos et al. (2015) reported that lipid-based nanoformulations improved systemic motility and bioefficacy of fungicides in palm oil plantations [9]. This approach not only increases the effectiveness of active compounds but also reduces environmental risks and potential for resistance development. This system can be classified into microemulsions and nanoemulsions, which differ fundamentally in their stability and formulation requirements.

Microemulsions are thermodynamically stable systems that form spontaneously and require high concentrations of surfactants and co-surfactants [10]. Although they can produce very small droplet sizes, the high surfactant content may raise concerns related to cost, phytotoxicity, and environmental impact, particularly in large-scale agricultural applications. In contrast, nanoemulsions are kinetically stable dispersions typically produced using high-energy methods such as high-shear homogenization or ultrasonication [10,11]. They can achieve comparable droplet sizes with significantly lower surfactant concentrations. High-energy emulsification method also allows better control over droplet size distribution and formulation composition. This makes them more suitable for sustainable agrochemical delivery.

Based on these considerations, this study aimed to develop a potent and an environmentally friendly hexaconazole-loaded nanoemulsion using clove oil as the dispersed phase for the control of *G. boninense*. Clove oil, which is rich in phenolic compounds such as eugenol, represents a promising lipid phase for the nanoemulsion system due to its intrinsic antifungal activity and favourable interfacial behaviour [12,13]. Incorporation of hexaconazole into a clove oil-based nanoemulsion may provide synergistic effects. These effects include enhanced fungicide solubilization, improved interfacial adhesion, and sustained release at the target site. As a result, effective antifungal inhibition may be achieved at lower active ingredient concentrations compared to conventional formulation.

In this study, the nanoemulsion was optimized through controlled variation of surfactant and co-stabiliser compositions. Its physicochemical properties were thoroughly characterized. By integrating colloidal science with biological performance evaluation, this work provides insight into the design of stable and eco-friendly nanofungicide systems

https://doi.org/10.3390/colloids10020024

for managing BSR while addressing environmental and health concerns associated with traditional fungicide use.

2. Materials and Methods

2.1. Materials

Hexaconazole (95%) was purchased from Haihaing Industry (Jinan, China), non-ionic tween series were purchased from Sigma-Aldrich (Darmstadt, Germany), sucrose ester was a gift from Sisterna (Roosendaal, The Netherlands), while phosphate-buffer saline (PBS), clove oil and potato dextrose agar (PDA) were purchased from Merck (Darmstadt, Germany). Glycerol was purchased from J.T. Baker (Phillipsburg, NJ, USA) and solvents including acetonitrile and methanol were purchased from Fisher Scientific (Waltham, MA, USA). *Ganoderma boninense* were provided by the Faculty of Agriculture, Universiti Putra Malaysia.

2.2. Selection of Nanoemulsion Compositions and Process Parameters

Preliminary screening was conducted to identify suitable surfactants and stabilizing agents for the development of a stable nanoemulsion with minimal particle size and polydispersity index (PDI). The formulations were prepared using a high-energy emulsification technique with different Tween series surfactants (5% *w*/*w*), namely Tween 20, Tween 40, Tween 60, Tween 80, and Tween 85, while maintaining constant concentrations of clove oil (5% *w*/*w*), glycerol (2% *w*/*w*). Hexaconazole concentration was fixed at 1% (*w*/*w*) for all screening and optimization experiments. Due to its hydrophobic nature, hexaconazole was first dissolved in the oil phase (clove oil) prior to emulsification to ensure uniform fungicide incorporation. Nanoemulsions were prepared by separately stirring oil and aqueous phase at 300 rpm until all components were completely dissolved. The oil phase was then gradually added to the aqueous phase and homogenized at 3000 rpm for 2 min using a high shear homogenizer (PT3100, Kinematica AG, Luzern, LU, Switzerland). The pre-mixed emulsion was then ultrasonicated (Q500, Qsonica, Newtown, CT, USA) for 1.5 min at 30% amplitude using intermittent mode (15 s on, 5 s off). In parallel, the effect of different stabilizing agents (2% *w*/*w*), including guar gum, xanthan gum, polyethylene glycol (PEG), pluronic F127 (F127), poloxamer 188 (P188), and sucrose ester (SE), was evaluated under identical formulation compositions and processing conditions. The prepared nanoemulsions were characterized for particle size and PDI using dynamic light scattering (DLS) as described in Section 2.3.

The surfactant and stabilizer that produced the smallest particle size and lowest polydispersity were selected for further formulation optimization. The One-Factor-At-a-Time (OFAT) method was used to optimize the nanomulsion compositions' concentration [14]. Formulations with varying concentrations of clove oil (2–10%, *w*/*w*), Tween 60 (2–10%, *w*/*w*), sucrose ester (1–5%, *w*/*w*) and glycerol (1–3%, *w*/*w*) were prepared. The optimized formulation composition was then subjected to ultrasonication at different durations (1–5 min) and amplitudes (10–50%). The optimal sonication time and amplitude were determined based on attaining the smallest particle size with minimal energy input and stable.

2.3. Particle Size Distribution and ζ-Potential Determination

The mean particle size, polydispersity index (PDI), and ζ-potential of the three independent sample were analyzed using a Zetasizer (Malvern Instruments, Worcestershire, UK) based on Dynamic Light Scattering (DLS) as described by Asmawi et al. [15]. To reduce multiple scattering effects, sample was diluted 1000-fold using deionized water.

https://doi.org/10.3390/colloids10020024

2.4. Viscosity Measurement

The viscosity of the Hexa-NE was determined using a rotational digital viscometer (Brookfield DV-II+, Middleboro, MA, USA) with a spindle SC4-18 at 25 ± 1 °C. The viscosity was measured in centipoise (cP), relied on a rotary transducer that gauged the deflection of the calibrated spring. The measurements were repeated in triplicate and the average viscosity value was recorded.

2.5. pH Measurement

The pH of the Hexa-NE was determined using a calibrated digital pH meter (Delta 320, Mettler Toledo, Tokyo, Japan). To ensure accuracy, the glass probe was rinsed with deionized water and then directly immersed in the sample without dilution. The measurement was conducted using three independent formulation samples at room temperature (28 ± 1 °C), and the average pH value of the sample was recorded.

2.6. Scanning Electron Microscope (SEM) and Transmission Electron Microscope (TEM)

The morphology of the Hexa-NE was observed using an Inspect F50 field emission scanning electron microscope (FEI, Hillsboro, OR, USA) and a transmission electron microscope (Hitachi H7100, Tokyo, Japan) [16].

2.7. In-Vitro Release Studies

Phosphate-buffered saline (PBS; pH 5.5) supplemented with 0.2% (*w*/*v*) Tween 80 was used as the dissolution medium to investigate the in-vitro release of hexaconazole from the nanoemulsion using a dialysis membrane diffusion method [17]. The pH 5.5 medium was used to simulate the conditions of potato dextrose agar used in the in vitro antifungal assays. The release study was carried out using three independent formulation samples by loading Hexa-NE formulation equivalent to 2 mg of hexaconazole into cellulose membrane dialysis bag with molecular weight cutoff 12 kDa. The prepared samples were immersed in dissolution media at 37 ± 1 °C and stirred at 250 rpm. At specified time intervals, 1 mL aliquots were withdrawn and immediately replaced with an equal volume of fresh medium to maintain sink conditions. The hexaconazole concentration in the collected samples was analysed using HPLC method [17]. The release profile was then evaluated using several kinetic models, including zero-order, first-order, Higuchi, Hixson–Crowell, and Korsmeyer–Peppas models. The experimental release data were fitted to each model, and the corresponding plots were analyzed to determine the model that provided the highest correlation coefficient of determination (R^2).

2.8. In-Vitro Antifungal Activity Efficacy Studies

The in vitro antifungal efficacy of hexaconazole solution and the Hexa-NE formulation against *G. boninense* was assessed using the poisoned medium technique on potato dextrose agar (PDA) [18]. The PDA medium was supplemented with either pure hexaconazole or freshly prepared Hexa-NE at various concentrations of the active ingredient (0, 0.1, 0.5, 1, and 2 µM), while PDA without the active ingredient served as the control. A 7 mm mycelial plug taken from the actively growing edge of *G. boninense* cultures was aseptically placed at the center of each treated agar plate. The inoculated plates were incubated at 28 ± 2 °C, and radial mycelial growth was monitored daily for seven days (n = 5). The antifungal activity was quantified by calculating the percentage inhibition of mycelial radial growth relative to the control.

2.9. Statistical Analysis

One-way analysis of variance (ANOVA) was performed, followed by Tukey's post hoc multiple-comparison test using GraphPad Prism (version 10.4.1, San Diego, CA, USA)

https://doi.org/10.3390/colloids10020024

to examine the data and identify significant differences between the experimental groups. Statistical significance was established at $p < 0.05$.

3. Results and Discussion

3.1. Screening of Formulation Compositions

As shown in Figure 1, this study investigated the effects of different Tweens series surfactants (Tween 20, 40, 60, 80, and 85) as well as various stabilizing agents on the particle size and polydispersity index (PDI). The results indicate that Tween 60 produced the smallest particle size and the lowest PDI compared to the other surfactants. Similarly, among the stabilizing agents tested, sucrose ester (SE) resulted in the smallest particle size and lowest PDI value, making it the most effective stabilizer. Therefore, Tween 60 and sucrose ester were chosen to be used for subsequent experiments.

(a)
(b)

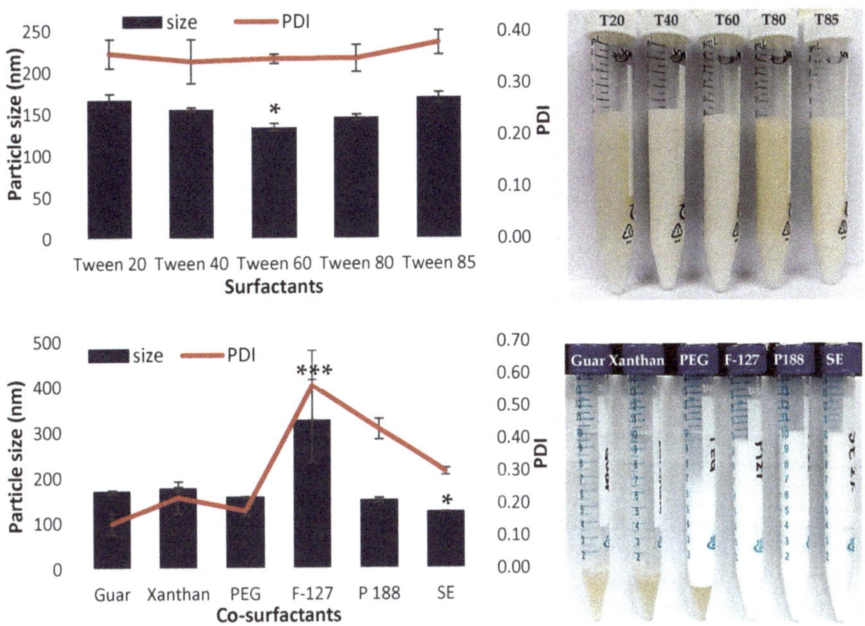

Figure 1. Effect of the nanoemulsion composition (**a**) surfactants; * $p < 0.05$ when compared to Tween 20; and (**b**) co-surfactants; * $p < 0.05$; *** $p < 0.001$ when compared to xanthan gum on the particle size and polydispersity index (PDI) value. Data presented as mean ± standard deviation, $n = 3$.

The One-Factor-At-a-Time (OFAT) approach was utilized to systematically determine the optimal concentration of each nanoemulsion component to achieve stable formulation with minimal particle size. Figure 2 presents the effects of varying the concentrations of different excipients, namely Tween 60, sucrose ester (SE), oil, and glycerol, on particle size of the nanoemulsion. The results indicated that increasing the concentrations of Tween 60, SE, and oil generally led to an increase in particle size. This effect may be attributed to the excess presence of surfactant and oil. Such excess can promote the formation of larger droplets or aggregates due to incomplete stabilization and possible coalescence when the interfacial area becomes oversaturated [19,20]. Conversely, at lower Tween 60 concentrations, insufficient surfactant molecules were available to fully surround the oil droplets. This led to poor stabilization and droplet aggregation due to incomplete interfacial coverage. However, variation in glycerol concentration did not significantly affect particle size. However, it plays an important role in modulating the osmotic balance and viscosity of the formulation [21]. Based on the optimization results, a nanoemulsion formulation known as Hexa-NE was developed. It contained 6% *w*/*w* clove oil, 2% *w*/*w* sucrose ester, 4% *w*/*w* Tween 60, 2% *w*/*w* glycerol, 1% *w*/*w* hexaconazole, and water q.s. to 100% *w*/*w*. This composition was selected to achieve an optimal balance between interfacial stabilization, droplet size reduction, and formulation stability.

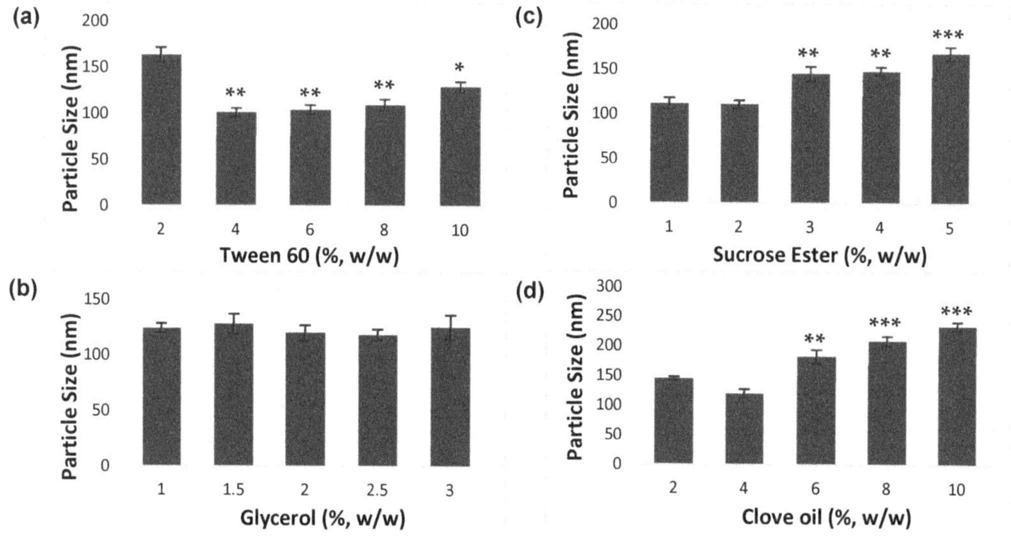

Figure 2. Effect of formulation components at different concentration of (**a**) Tween 60, (**b**) sucrose ester, (**c**) glycerol, and (**d**) clove oil on the particle size of Hexa-NE using the One-Factor-At-a-Time (OFAT) approach. *** $p < 0.001$; ** $p < 0.01$, * $p < 0.05$ when compared with the lowest concentration of each component. Data presented as mean ± standard deviation, $n = 3$.

The effect of processing parameters on particle size and distribution were further evaluated. In general, increasing sonication time reduced nanoemulsion particle size across all formulation (Figure 3). Similar observations were reported by Guttoff et al. [22] and Saberi et al. [21], whereby the particle size exhibits reduction with increased in homogenization energy. This confirms the effectiveness of ultrasonic homogenization in droplet size reduction. Larger droplets observed at shorter sonication times may be attributed to insufficient cavitation energy. Increasing sonication time enhances shear forces and promotes droplet breakup [23,24]. Although higher ultrasonic power can reduce processing time, excessive energy input may cause droplet recoalescence and localized heating. This effects may compromise nanoemulsion stability [23]. Therefore, an intermediate ultrasonic amplitude combined with an optimized sonication time was selected to achieve efficient droplet size reduction while minimizing thermal and mechanical stress. As shown in Figure 3, beyond 3 min of sonication at higher amplitude (40% and 50%), particle size reduction showed no further substantial decrease. This indicates an optimal sonication duration for efficient nanoemulsion formation. Accordingly, sonication time (3 min) and amplitude (30%) were selected based on achieving the smallest particle size with minimal energy input and without signs of instability.

https://doi.org/10.3390/colloids10020024

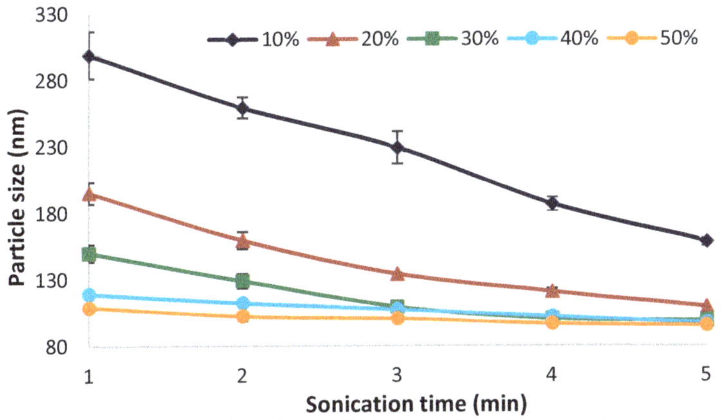

Figure 3. Effect of sonication time and amplitude for preparation of nanoemulsion using ultrasonicator homogenizer on the particle size of Hexa-NE.

3.2. Physicochemical and Characterization of Hexa-NE Nanoemulsion

The physicochemical characteristics of the Hexa-NE play important roles in defining its long-term stability and suitability as an efficient foliar fungicide delivery system. With a mean particle size of 105.8 ± 2.9 nm, the formulation falls within the optimal nanoscale range. This small particle size provides a high surface area that promotes strong adhesion to plant surfaces and improved permeability through the cuticle barrier. This finding is consistent with previous reports indicating that nanoemulsions with droplet sizes below 200 nm were able to significantly and effectively increase the foliar uptake of active ingredients [25]. Hexa-NE showed moderate polydispersity, as reflected by its polydispersity index (PDI) which is 0.358 ± 0.038. Generally, reduced particle size and narrower size distribution improve nanoemulsion stability and minimize Ostwald ripening. Although further reduction may be obtained through multivariate optimization strategies or higher-energy processing techniques, the moderate polydispersity observed in this study was sufficient to maintain physical stability. Hence, contributes to minimize droplet aggregation during storage [26]. The stability of Hexa-NE is further supported by its highly negative zeta potential (−53.53 ± 0.86 mV). A strong surface charge generates strong electrostatic repulsion between droplets, thus minimizing the likelihood of coalescence [27]. This aligns with the findings of Kumar et al. (2021), who reported that zeta potential values exceeding ±30 mV are critical for maintaining stable nanoemulsion systems [28].

Additionally, the viscosity of Hexa-NE was low (30.24 ± 1.06 mPa·s) and suitable for an efficient aerosolization while still allowing sufficient retention on leaf surfaces. The formulation also exhibited a lower surface tension (23.87 ± 1.86 mN/m) as compared to water. This reduction promotes enhanced wetting and spreading over hydrophobic plant cuticles. High viscosity and surface tension can reduce aerosolization efficiency, as greater energy is required to generate fine droplets. This may also contributes to an increase in aerodynamic diameter [29]. Improved spreading behavior has been directly linked to increased pesticide coverage and deposition, as previously demonstrated by Chen et al. [30]. Moreover, the formulation's pH of 6.14 ± 0.01 lies within a safe range for plant tissues. This minimizes the risk of phytotoxicity and ensures compatibility with foliar application practices [31].

Transmission Electron Microscopy (TEM) was utilized to determine the particle shape of nanoemulsions and the particle size and homogeneity measured using DLS was consistent with TEM analysis (Figure 4). The absence of aggregation and the well-defined droplet

https://doi.org/10.3390/colloids10020024

boundaries, further confirm the structural stability of the nanoemulsion. This observations further support its suitability for foliar application where consistency and stability are critical for performance.

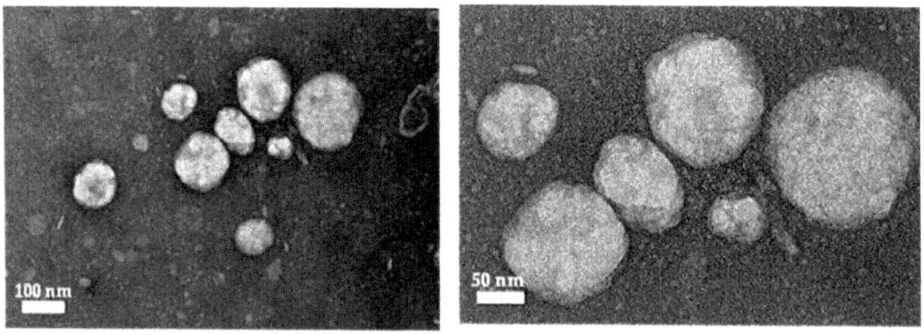

(a) (b)

Figure 4. The transmission electron microscope (TEM) images of Hexa-Ne with scale bar (**a**) 100 nm and (**b**) 50 nm.

3.3. *In Vitro Fungicide Release*

The release profiles of hexaconazole solution and Hexa-NE at pH 5.5 demonstrated different release behaviors. The hexaconazole solution exhibited rapid and nearly complete release within the initial 24 h, indicating immediate availability of the hexaconazole in the dissolution medium. While such an immediate release may provide a quick onset of antifungal action, it may also result a shorter duration of effectiveness and a higher likelihood of rapid depletion of the active ingredient. In contrast, Hexa-NE displayed a controlled and sustained release profile over 120 h. Although an initial burst release was observed during the early stage, likely due to the release of hexaconazole located at or near the oil-water interface. However, the overall release rate was substantially slower than that of the hexaconazole solution. The sustained release behavior of Hexa-NE can be attributed to the encapsulation of hexaconazole within the oil phase of the nanoemulsion droplets. Additional time is required for the hexaconazole to diffuse through the oil droplet core and the surrounding surfactant interfacial layer before reaching the dissolution medium, which consequently slows its release rate. Additionally, the presence of surfactant and stabilizer layers surrounding the droplets further modulates hexaconazole diffusion [20,32].

The mechanism for fungicide release from formulation was predicted by fitting the release data to zero-order, first-order, Higuchi, Hixson-Crowell and Korsmeyer-Peppas models. In this study, linear regression analysis was used to determine correlation coefficients (R^2) for each model. As shown in Figure 5, among the tested models, the first-order kinetic model provided the best fit for Hexa-NE formulation having the highest R^2 value. This finding indicates that hexaconazole release from Hexa-NE is governed predominantly by concentration-driven diffusion from the oil core, modulated by interfacial and droplet-size effects. The absence of an excessive burst effect and the gradual increase in cumulative release over 120 h further support the controlled and sustained release behavior. This release profile of Hexa-NE may contribute to prolonged antifungal exposure at the target site, which is consistent with the enhanced in vitro growth inhibition observed at low concentrations. However, further *in planta* and field evaluations are required to confirm whether this translates into extended protection under agricultural conditions.

https://doi.org/10.3390/colloids10020024

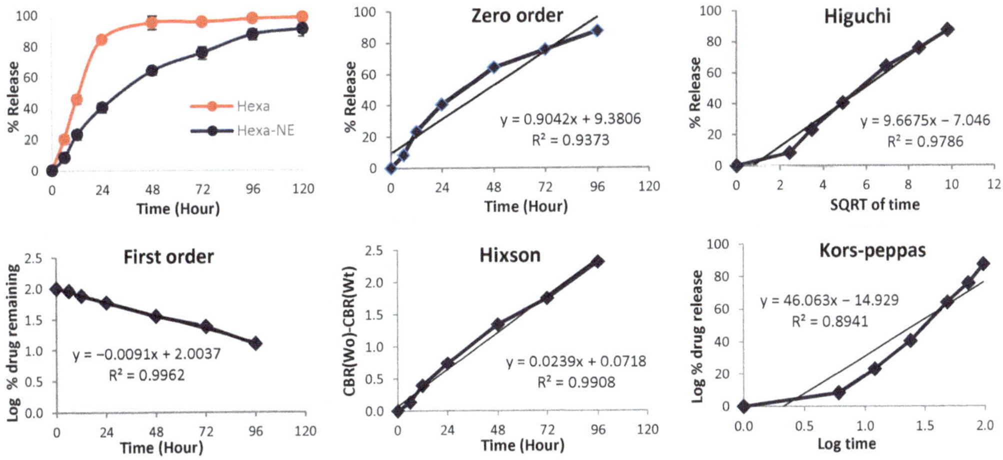

Figure 5. The release profiles of hexaconazole from Hexa-NE at pH 5.5 and kinetic model fitting of the release data using zero-order, first-order, Higuchi, Hixson-Crowell, and Korsmeyer-Peppas models.

3.4. In Vitro Antifungal Activity Assay on G.boninense

The antifungal efficacy of hexaconazole and Hexa-NE against *G. boninense* was tested through agar plate assays, as shown in Figure 6a. The untreated agar plate (control) showed aggressive and dense growth of *G. boninense* mycelial almost filling the plate after seven days of incubation. However, agar plate containing 0.1 µM hexaconazole solution showed reduced fungal growth. This can be seen through the smaller colony size compared to the control. These observations confirm that hexaconazole has antifungal activity against *G. boninense*, which works by inhibiting the biosynthesis of ergosterol, an important component of fungal cell membranes [33]. Interestingly, agar plate containing same concentration of Hexa-NE (0.1 µM) exhibited the most significant inhibition of fungal growth compared to hexaconazole solution, producing a very small and compact colony.

In addition, SEM analysis was also conducted to obtain more information about the mechanism behind the increased antifungal activity of Hexa-NE. As shown in Figure 6b, the untreated hyphae of *G. boninense* showed healthy fungal growth and normal morphology with branching filamentous, smooth tubular surfaces, and rigid cell walls. Conversely, hyphae treated with hexaconazole solution showed moderate structural alterations, including uneven surfaces, tubular shrinkage, and partial deformation. These structural changes are consistent with disturbance in membrane integrity and cell wall synthesis affected by ergosterol depletion [3,34]. Nevertheless, higher frequency of morphological hyphae damage was observed in the Hexa-NE treated sample with concentration of 0.1 µM. The hyphae appear severely distorted, with thinning, wrinkling and breakage. This severe damage indicates loss of membrane integrity and cytoplasmic leakage, which results in the death fungal cell [35]. The enhanced antifungal activity of Hexa-NE may be attributed to improved penetration of hexaconazole across the fungal cell wall. This might be due to its nanosized and lipid-based properties as it can interact strongly with ergosterol-rich membrane and improves the fusion into the fungal cell [36].

To further confirm the antifungal effectiveness as observed on the plate assay and SEM analysis, growth inhibition of *G. boninense* was further evaluated using different concentrations of hexaconazole and Hexa-NE (Figure 7). The results displayed a concentration-dependent inhibition pattern for both samples. However, Hexa-NE consistently exhibits high antifungal activity compared to hexaconazole solution, especially at lower concentrations. At 0.1 μM, hexaconazole solution showed moderate inhibitory activity, reaching approximately 40% growth inhibition. Conversely, Hexa-NE at the same concentration produced an almost total growth inhibition (97.1%), exhibiting a significant increase in antifungal potential when nanoemulsified. This observation strongly supports the very small and compact colony size, and significant hyphae deformation seen earlier in Figure 6b.

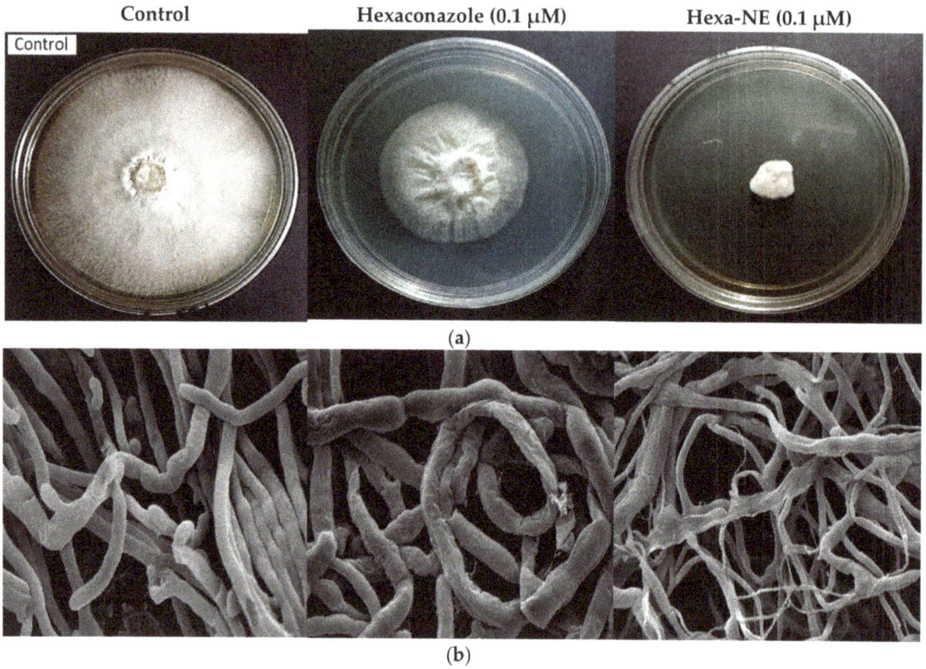

Figure 6. The antifungal activity against *G. boninense* after seven days of incubation at 28 ± 2 °C. (**a**) Images of *G. boninense* on potato dextrose agar plates and (**b**) SEM images of hyphae of *G.boninense* untreated (control) and after treatment with hexaconazole solution and Hexa-NE at 0.1 μM concentration.

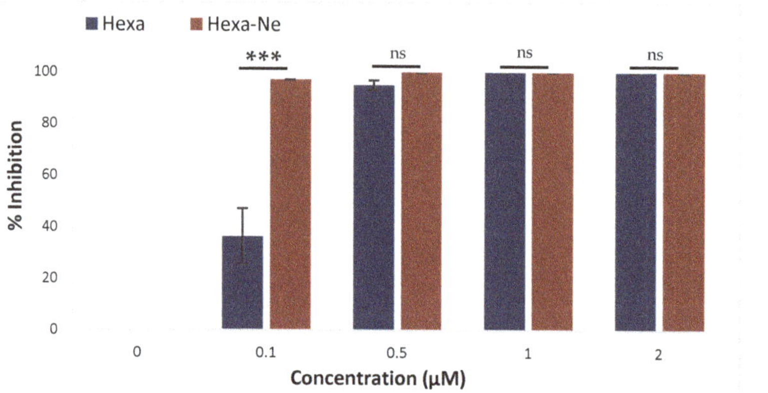

Figure 7. Antifungal effect on the growth inhibition of *G. boninense* by Hexaconazole solution (Hexa) and Hexaconazole-loaded nanoemulsion (Hexa-NE) after seven days incubation at 28 ± 2 °C. Data presented as mean ± standard deviation, $n = 3$. *** $p < 0.001$ when compared to 0.1 μM Hexa-NE and ns represents non-significant.

https://doi.org/10.3390/colloids10020024

These findings indicate that Hexa-NE can achieve enhanced antifungal activity at lower active ingredient concentrations compared to the conventional hexaconazole solution. This may allow reduction of fungicide dosage, which may reduce overall chemical input. Nevertheless, further investigations including ecotoxicity, effects on non-target organisms, and phytotoxicity assessments are required to confirm whether such dose reduction translates into measurable environmental and health benefits. In comparison with previously reported hexaconazole-loaded nanoparticles [18,37,38], which demonstrated improved stability and antifungal performance through polymeric encapsulation, the present Hexa-NE system offers a lipid-based delivery approach incorporating clove oil as a functional dispersed phase without complex crosslinking steps. Besides, the sustained release behavior observed in Hexa-NE, together with its strong in vitro efficacy at low concentration, highlights an alternative formulation strategy that may offer practical advantages in terms of preparation simplicity and potential scalability for agricultural application.

4. Conclusions

This study successfully developed a stable hexaconazole-loaded nanoemulsion (Hexa-NE) with optimized formulation and process parameters using OFAT method. The developed Hexa-NE produced nano-scale particle size and exhibited physicochemical properties suitable for foliar application. This has been proven by the significant antifungal effectiveness against *G. boninense* compared to hexaconazole solution. In addition, Hexa-NE achieved almost complete inhibition of fungal growth at low concentrations and caused severe hyphae damage as shown in SEM analysis. This indicates Hexa-NE's ability to penetrate better into the fungal and disrupt its membranes. These findings also suggest that nanoemulsification able to increase dose efficiency and anti-fungal performance that has more sustainable potential to manage basal stem rot disease in palm oil. Nevertheless, this study is limited to in vitro antifungal evaluation and single-factor formulation optimization. Future research should incorporate multivariate experimental designs, field-scale validation, long-term storage stability, and comprehensive environmental impact studies to fully establish the applicability of Hexa-NE in oil palm plantations.

Author Contributions: Conceptualization, A.A.A., M.B.A.R. and N.A.I.M.Z.; methodology, A.A.A., M.B.A.R. and N.A.I.M.Z.; software, A.A.A.; validation, A.A.A.; formal analysis, A.A.A., N.F.A.A. and M.F.A.; investigation, A.A.A., N.F.A.A. and M.F.A.; resources, M.B.A.R., N.A.I.M.Z., N.A.M.A. and F.A.; data curation, A.A.A. and M.B.A.R.; writing—original draft preparation, A.A.A.; writing—review and editing, M.B.A.R., N.A.M.A. and F.A.; visualization, A.A.A.; supervision, M.B.A.R., N.A.I.M.Z., N.A.M.A. and F.A.; project administration, A.A.A., M.B.A.R. and N.A.M.A.; funding acquisition, A.A.A. and M.B.A.R. All authors have read and agreed to the published version of the manuscript.

Funding: This research was funded by the Kurita Water and Environment Foundation (KWEF), grant number 22Pmy249-R3.

Data Availability Statement: All data relevant to the publication are included.

Acknowledgments: Sincere appreciation to Universiti Putra Malaysia (UPM) and Universiti Malaysia Pahang Al-Sultan Abdullah (UMPSA) for the facilities provided throughout this research. The authors take full responsibility for the content of this publication. English editing, including grammar and sentence structure improvement, was assisted using Grammarly. All authors have reviewed and approved the final manuscript.

Conflicts of Interest: The authors declare no conflict of interest. The funders had no role in the design of the study; in the collection, analyses, or interpretation of data; in the writing of the manuscript; or in the decision to publish the results.

https://doi.org/10.3390/colloids10020024

Abbreviations

The following abbreviations are used in this manuscript:

TEM	Transmission Electron Microscopy
SEM	Scanning Electron Microscopy
PDI	Polydispersity index
Hexa	Hexaconazole
Hexa-NE	Hexaconazole nanoemulsion
PBS	Phosphate-buffer saline
PEG	Polyethylene glycol
SE	Sucrose Ester
F-127	Pluronic F-128
P188	Poloxamer 188

References

1. Statista. Production Volume of Palm oil Worldwide from 2000/2001 to 2025/2026. Available online: https://www.statista.com/statistics/613471/palm-oil-production-volume-worldwide/ (accessed on 10 October 2023).
2. Cheah, W.Y.; Siti-Dina, R.P.; Leng, S.T.K.; Er, A.C.; Show, P.L. Circular Bioeconomy in Palm Oil Industry: Current Practices and Future Perspectives. *Environ. Technol. Innov.* **2023**, *30*, 103050. [CrossRef]
3. Zakaria, L. Basal Stem Rot of Oil Palm: The Pathogen, Disease Incidence, and Control Methods. *Plant Dis.* **2023**, *107*, 603–615. [CrossRef]
4. Asmawi, A.A.; Adam, F.; Mohd Azman, N.A.; Abdul Rahman, M.B. Advancements in the Nanodelivery of Azole-Based Fungicides to Control Oil Palm Pathogenic Fungi. *Heliyon* **2024**, *10*, e37132. [CrossRef]
5. Maluin, F.N.; Hussein, M.Z.; Idris, A.S. An Overview of the Oil Palm Industry: Challenges and Some Emerging Opportunities for Nanotechnology Development. *Agronomy* **2020**, *10*, 356. [CrossRef]
6. Hazra, R.S.; Roy, J.; Jiang, L.; Webster, D.C.; Rahman, M.M.; Quadir, M. Biobased, Macro-, and Nanoscale Fungicide Delivery Approaches for Plant Fungi Control. *ACS Appl. Bio Mater.* **2023**, *6*, 2698–2711. [CrossRef]
7. Kumar, S.; Nehra, M.; Dilbaghi, N.; Marrazza, G.; Hassan, A.A.; Kim, K.H. Nano-Based Smart Pesticide Formulations: Emerging Opportunities for Agriculture. *J. Control. Release* **2019**, *294*, 131–153. [CrossRef]
8. Mosa, M.A.; Youssef, K.; Hamed, S.F.; Hashim, A.F. Antifungal Activity of Eco-Safe Nanoemulsions Based on *Nigella Sativa* Oil against *Penicillium Verrucosum* Infecting Maize Seeds: Biochemical and Physiological Traits. *Front. Microbiol.* **2023**, *13*, 1108733. [CrossRef] [PubMed]
9. Campos, E.V.R.; De Oliveira, J.L.; Da Silva, C.M.G.; Pascoli, M.; Pasquoto, T.; Lima, R.; Abhilash, P.C.; Fernandes Fraceto, L. Polymeric and Solid Lipid Nanoparticles for Sustained Release of Carbendazim and Tebuconazole in Agricultural Applications. *Sci. Rep.* **2015**, *5*, 13809. [CrossRef]
10. Magrode, N.; Poomanee, W.; Kiattisin, K.; Ampasavate, C. Microemulsions and Nanoemulsions for Topical Delivery of Tripeptide-3: From Design of Experiment to Anti-Sebum Efficacy on Facial Skin. *Pharmaceutics* **2024**, *16*, 554. [CrossRef]
11. Souto, E.B.; Cano, A.; Martins-Gomes, C.; Coutinho, T.E.; Zielińska, A.; Silva, A.M. Microemulsions and Nanoemulsions in Skin Drug Delivery. *Bioengineering* **2022**, *9*, 158. [CrossRef] [PubMed]
12. Hashem, A.H.; Abdelaziz, A.M.; Hassanin, M.M.H.; Al-Askar, A.A.; AbdElgawad, H.; Attia, M.S. Potential Impacts of Clove Essential Oil Nanoemulsion as Bio Fungicides against Neoscytalidium Blight Disease of *Carum carvi* L. *Agronomy* **2023**, *13*, 1114. [CrossRef]
13. Ahmad, I.; Farheen, M.; Kukreti, A.; Afzal, O.; Akhter, M.H.; Chitme, H.; Visht, S.; Altamimi, A.S.A.; Alossaimi, M.A.; Alsulami, E.R.; et al. Natural Oils Enhance the Topical Delivery of Ketoconazole by Nanoemulgel for Fungal Infections. *ACS Omega* **2023**, *8*, 28233–28248. [CrossRef] [PubMed]
14. Ngan, C.L.; Fard Masoumi, H.R.; Basri, M.; Abdul Rahman, M.B. Development of Nano-Colloidal System for Fullerene by Ultrasonic-Assisted Emulsification Techniques Based on Artificial Neural Network. *Arab. J. Chem.* **2016**, *12*, 4162–4170. [CrossRef]
15. Asmawi, A.A.; Salim, N.; Abdulmalek, E.; Abdul Rahman, M.B. Size-Controlled Preparation of Docetaxel- and Curcumin-Loaded Nanoemulsions for Potential Pulmonary Delivery. *Pharmaceutics* **2023**, *15*, 652. [CrossRef] [PubMed]
16. Asmawi, A.A.; Salim, N.; Ngan, C.L.; Ahmad, H.; Abdulmalek, E.; Masarudin, M.J.; Abdul Rahman, M.B. Excipient Selection and Aerodynamic Characterization of Nebulized Lipid-Based Nanoemulsion Loaded with Docetaxel for Lung Cancer Treatment. *Drug Deliv. Transl. Res.* **2018**, *9*, 543–554. [CrossRef]

https://doi.org/10.3390/colloids10020024

17. Ahmad Aljafree, N.F.; Ahmad, M.F.; Abd Aziz, U.; Borzehandani, M.Y.; Mohamad Jaafar, A.; Asib, N.; Nguyen, H.L.; Mohamed Tahir, M.I.; Mohammad Latif, M.A.; Cordova, K.E.; et al. Calcium L-Malate and d-Tartarate Frameworks as Adjuvants for the Sustainable Delivery of a Fungicide. *ACS Appl. Mater. Interfaces* **2025**, *17*, 17672–17683. [CrossRef]
18. Maluin, F.N.; Hussein, M.Z.; Yusof, N.A.; Fakurazi, S.; Idris, A.S.; Hilmi, N.H.Z.; Daim, L.D.J. Preparation of Chitosan-Hexaconazole Nanoparticles as Fungicide Nanodelivery System for Combating Ganoderma Disease in Oil Palm. *Molecules* **2019**, *24*, 2498. [CrossRef]
19. Mushtaq, A.; Mohd Wani, S.; Malik, A.R.; Gull, A.; Ramniwas, S.; Ahmad Nayik, G.; Ercisli, S.; Alina Marc, R.; Ullah, R.; Bari, A. Recent Insights into Nanoemulsions: Their Preparation, Properties and Applications. *Food Chem. X* **2023**, *18*, 100684. [CrossRef]
20. Sarheed, O.; Dibi, M.; Ramesh, K.V.R.N.S. Studies on the Effect of Oil and Surfactant on the Formation of Alginate-Based O/W Lidocaine Nanocarriers Using Nanoemulsion Template. *Pharmaceutics* **2020**, *12*, 1223. [CrossRef]
21. Saberi, A.H.; Fang, Y.; McClements, D.J. Effect of Glycerol on Formation, Stability, and Properties of Vitamin-E Enriched Nanoemulsions Produced Using Spontaneous Emulsification. *J. Colloid Interface Sci.* **2013**, *411*, 105–113. [CrossRef]
22. Guttoff, M.; Saberi, A.H.; McClements, D.J. Formation of Vitamin D Nanoemulsion-Based Delivery Systems by Spontaneous Emulsification: Factors Affecting Particle Size and Stability. *Food Chem.* **2015**, *171*, 117–122. [CrossRef]
23. Li, W.; Leong, T.S.H.; Ashokkumar, M.; Martin, G.J.O. A Study of the Effectiveness and Energy Efficiency of Ultrasonic Emulsification. *Phys. Chem. Chem. Phys.* **2017**, *20*, 86–96. [CrossRef]
24. Ascrizzi, G.I.; Fuenmayor, C.A.; Piazza, L. Ultrasound-Assisted Nanoemulgel Preparation: A One-Step Approach for Enhanced Rheo-Tribological Properties. *Innov. Food Sci. Emerg. Technol.* **2025**, *106*, 104246. [CrossRef]
25. Yanasan, N.; Wangkananon, W.; Natakankitkul, S.; Kiattisin, K. Nanoemulsions Containing *Passiflora Quadrangularis* L. Fruit Extracts for Cosmetic Application and Skin Efficacy Study. *Cosmet* **2024**, *11*, 57. [CrossRef]
26. Moreira, J.B.; Goularte, P.G.; Morais, M.G.; Costa, J.A.V. Preparation of Beta-Carotene Nanoemulsion and Evaluation of Stability at a Long Storage Period. *Food Sci. Technol.* **2019**, *39*, 599–604. [CrossRef]
27. Kurpiers, M.; Wolf, J.D.; Steinbring, C.; Zaichik, S.; Bernkop-Schnürch, A. Zeta Potential Changing Nanoemulsions Based on Phosphate Moiety Cleavage of a PEGylated Surfactant. *J. Mol. Liq.* **2020**, *316*, 113868. [CrossRef]
28. Kumar, A.; Gradzielski, M.; Kanwar, R.; Mehta, S.K. Optimization and Stability of Geraniol Nanoemulsion Stabilized Synergistically Using Mixed Biosurfactant System of Saponin and Lecithin. *Langmuir* **2025**, *41*, 10108–10121. [CrossRef]
29. Nasr, M.; Nawaz, S.; Elhissi, A. Amphotericin B Lipid Nanoemulsion Aerosols for Targeting Peripheral Respiratory Airways via Nebulization. *Int. J. Pharm.* **2012**, *436*, 611–616. [CrossRef] [PubMed]
30. Chen, Y.F.; Yang, C.H.; Chang, M.S.; Ciou, Y.P.; Huang, Y.C. Foam Properties and Detergent Abilities of the Saponins from Camellia Oleifera. *Int. J. Mol. Sci.* **2010**, *11*, 4417–4425. [CrossRef]
31. Carrasco-Sandoval, J.; Aranda-Bustos, M.; Henríquez-Aedo, K.; López-Rubio, A.; Fabra, M.J. Bioaccessibility of Different Types of Phenolic Compounds Co-Encapsulated in Alginate/Chitosan-Coated Zein Nanoparticles. *LWT* **2021**, *149*, 112024. [CrossRef]
32. Wilson, R.J.; Li, Y.; Yang, G.; Zhao, C.X. Nanoemulsions for Drug Delivery. *Particuology* **2022**, *64*, 85–97. [CrossRef]
33. Roy, I.; Thapa, M.; Goswami, A. Nanohexaconazole: Synthesis, Characterisation and Efficacy of a Novel Fungicidal Nanodispersion. *IET Nanobiotechnol.* **2018**, *12*, 864. [CrossRef]
34. Haw, Y.H.; Lai, K.W.; Chuah, J.H.; Bejo, S.K.; Husin, N.A.; Hum, Y.C.; Yee, P.L.; Tee, C.A.T.H.; Ye, X.; Wu, X. Classification of Basal Stem Rot Using Deep Learning: A Review of Digital Data Collection and Palm Disease Classification Methods. *PeerJ Comput. Sci.* **2023**, *9*, 1–30. [CrossRef] [PubMed]
35. Zhai, S.; Guo, H.; Sun, T.; Chen, J.; Guo, M.; Chen, G. 3-Methyl-1-Butanol Inhibited Gray Mold of Red Grape by Damaging Cell Membrane Integrity and the Antioxidant Capacity of Botrytis Cinerea under Oxidative Stress. *LWT* **2025**, *231*, 118328. [CrossRef]
36. Ermakova, E.; Zuev, Y. Effect of Ergosterol on the Fungal Membrane Properties. All-Atom and Coarse-Grained Molecular Dynamics Study. *Chem. Phys. Lipids* **2017**, *209*, 45–53. [CrossRef]
37. Mustafa, I.F.; Hussein, M.Z. Synthesis and Technology of Nanoemulsion-Based Pesticide Formulation. *Nanomaterials* **2020**, *10*, 1608. [CrossRef] [PubMed]
38. Gao, S.; Guo, X.; Li, F.; Zhang, Y.; Yu, Y.; Fu, Y.; Ye, F. Delivery of Hexaconazole by the Carboxymethylcellulose Grafting and Cellulase/PH Responsiveness Hollow Mesoporous Silica Nanoparticles. *Carbohydr. Polym.* **2025**, *365*, 123822. [CrossRef] [PubMed]

https://doi.org/10.3390/colloids10020024

MDPI AG
Grosspeteranlage 5
4052 Basel
Switzerland
Tel.: +41 61 683 77 34

Colloids and Interfaces Editorial Office
E-mail: colloids@mdpi.com
www.mdpi.com/journal/colloids